Cambridge Elements

Elements in Climate Change and Cities: Third Assessment Report of the Urban Climate Change Research Network

Series Editors

William Solecki
New York

Minal Pathak
Ahmedabad

Martha Barata
Rio de Janeiro

Aliyu Salisu Barau
Kano

Maria Dombrov
New York

Cynthia Rosenzweig
New York

LEARNING FROM COVID-19 FOR CLIMATE-READY URBAN TRANSFORMATION

Coordinating Lead Authors

Darshini Mahadevia
Ahmedabad

Gian C. Delgado Ramos
Mexico City

Lead Authors

Janice Barnes
New York

Joan Fitzgerald
Boston

Miho Kamei
Hayama

Kevin Lanza
Austin

Shaftesbury Road, Cambridge CB2 8EA, United Kingdom

One Liberty Plaza, 20th Floor, New York, NY 10006, USA

477 Williamstown Road, Port Melbourne, VIC 3207, Australia

314–321, 3rd Floor, Plot 3, Splendor Forum, Jasola District Centre, New Delhi – 110025, India

103 Penang Road, #05–06/07, Visioncrest Commercial, Singapore 238467

Cambridge University Press is part of Cambridge University Press & Assessment, a department of the University of Cambridge.

We share the University's mission to contribute to society through the pursuit of education, learning and research at the highest international levels of excellence.

www.cambridge.org
Information on this title: www.cambridge.org/9781009527255

DOI: 10.1017/9781009527279

© Darshini Mahadevia, Gian C. Delgado Ramos, Janice Barnes, Joan Fitzgerald, Miho Kamei, Kevin Lanza and contributors 2025

This publication is in copyright. Subject to statutory exception and to the provisions of relevant collective licensing agreements, with the exception of the Creative Commons version the link for which is provided below, no reproduction of any part may take place without the written permission of Cambridge University Press & Assessment.

An online version of this work is published at doi.org/10.1017/9781009527279 under a Creative Commons Open Access license CC-BY-NC 4.0 which permits re-use, distribution and reproduction in any medium for non-commercial purposes providing appropriate credit to the original work is given and any changes made are indicated. To view a copy of this license visit https://creativecommons.org/licenses/by-nc/4.0

When citing this work, please include a reference to the DOI 10.1017/9781009527279

First published 2025

A catalogue record for this publication is available from the British Library

ISBN 978-1-009-52725-5 Hardback
ISBN 978-1-009-52729-3 Paperback
ISSN 2976-9116 (online)
ISSN 2976-9108 (print)

Additional resources for this publication at www.cambridge.org/mahadevia

Cambridge University Press & Assessment has no responsibility for the persistence or accuracy of URLs for external or third-party internet websites referred to in this publication and does not guarantee that any content on such websites is, or will remain, accurate or appropriate.

Learning from COVID-19 for Climate-Ready Urban Transformation

Elements in Climate Change and Cities: Third Assessment Report of the Urban Climate Change Research Network

DOI: 10.1017/9781009527279
First published online: February 2025

Coordinating Lead Authors
Darshini Mahadevia
Ahmedabad

Gian C. Delgado Ramos
Mexico City

Lead Authors
Janice Barnes
New York

Joan Fitzgerald
Boston

Miho Kamei
Hayama

Kevin Lanza
Austin

Author for correspondence: Darshini Mahadevia
darshini.mahadevia@ahduni.edu.in

Abstract: Cities have suffered from the COVID-19 pandemic and are increasingly experiencing exacerbated heatwaves, floods, and droughts due to climate change. Going forward, cities need to address both climate and public health crises effectively while reducing poverty and inequity, often in the context of economic pressure and declining levels of trust in government. The COVID-19 pandemic has revealed gaps in city readiness for simultaneous responses to pandemics and climate change, particularly in the Global South. However, these concurrent challenges to cities present an opportunity to reformulate current urbanization patterns and the economies and dynamics they enable. This Element focuses on understanding how COVID-19 and city-level responses impacted climate change mitigation and adaptation, and vice versa, in terms of warnings, lessons learned, and calls to action. This title is also available as open access on Cambridge Core.

Keywords: COVID-19, climate change, adaptation, resilience, mitigation, urban transformation

© Darshini Mahadevia, Gian C. Delgado Ramos, Janice Barnes, Joan Fitzgerald, Miho Kamei, Kevin Lanza and contributors 2025

ISBNs: 9781009527255 (HB), 9781009527293 (PB), 9781009527279 (OC)
ISSNs: 2976-9116 (online), 2976-9108 (print)

Contents

List of Contributors — v

Series Preface — 1

Foreword I – Dr. Daniel Soranz, Municipal Health Secretary of Rio de Janeiro, Brazil — 9

Foreword II – Lykke Leonardsen, Programme Director, Resilient and Sustainable City Solutions at the City of Copenhagen, Denmark — 10

Series Editors' Introduction to *Learning from COVID-19 for Climate-Ready Urban Transformation* — 11

Major Findings and Key Messages — 12

1 Introduction and Framing — 15

2 Vulnerability and Informality — 19

3 Interactions at the Urban Scale — 28

4 Urban Systems: Built Environment, Transportation, and Waste — 38

5 Interactions with Urban Ecology — 48

6 Governance and Urban Climate Action — 50

7 Energy and Economics — 56

8 Learning from COVID-19: Accelerating Urban Transformational Pathways — 61

Appendix: UCCRN ARC3.3 Stakeholder Soundings — 68

References — 71

Contributors

Contributing Authors / Case Study Authors[*]

Zaheer Allam
Geelong

Amita Bhide
Mumbai

Yakubu Bununu
Zaria

Didier Chabaud
Paris

Amitkumar Dubey
Ahmedabad

Yann Francoise
Paris

Saumya Lathia
Ahmedabad

María Fernanda Mac Gregor-Gaona
Mexico City

Carlos Moreno
Tunja/Paris

Marie-Christine Therrien
Montreal

Nada Toueir
Montreal

Nelzair Vianna
Salvador

Element Scientist[†]

Melissa López Portillo-Purata
Mexico City

---

[*] UCCRN *Assessment Reports on Climate Change and Cities* (ARC3) authors are associated solely with their cities or metropolitan regions. Suggested Citation: Mahadevia, D., G.C. Delgado Ramos, J. Barnes, J. Fitzgerald, M. Kamei, K. Lanza. 2025. Learning from COVID-19 for Climate-Ready Urban Transformation. In Solecki, W., M. Pathak, M. Barata, A. Barau, M. Dombrov, and C. Rosenzweig (Eds.), Climate Change and Cities: Third Assessment Report of the Urban Climate Change Research Network. Cambridge: Cambridge University Press.

[†] ARC3 Element Scientists support the Coordinating Lead Authors in content development and research support.

Learning from COVID-19

Series Preface

Urban Climate Change Research Network

Third Assessment Report on Climate Change and Cities (ARC3.3)

William Solecki (New York), Minal Pathak (Ahmedabad), Martha Barata (Rio de Janeiro), Aliyu Salisu Barau (Kano), Maria Dombrov (New York), and Cynthia Rosenzweig (New York)

Cities and the urbanization process itself are at a crossroads. While the world's urban population continues to grow, cities are increasingly pressed by chronic and acute stresses like increasing inequity, polluted air and waters, limited governance and financial capacities, along with entrenched spasmodic crime and conflict – and the COVID-19 pandemic. Climate change has now exacerbated these problems and in many cases created new ones, at a time when cities are asked to be the bulwark of the climate solution space. The advent and application of new technologies associated with the internet, environmental sensing, multimodal transport, and planning and design strategies portend a new golden age of sustainable cities. Some cities provide glimmers of this possible future, but persistent stresses and crises, along with climate change, push against progress. In the Urban Climate Change Research Network's (UCCRN) *Third Assessment Report on Climate Change and Cities* (ARC3.3), we directly address these issues head-on and present state-of-the-art knowledge on how to bring all cities and their residents forward to a more sustainable future.

An absolute necessity now exists for all cities, both in the Global North and the Global South, to work aggressively to fulfill their potential as leaders in climate change action. In the Global North, the task is for cities to address the emerging challenges from the changing climate and the exigencies of compliance with the United Nations Framework Convention on Climate Change (UNFCCC) Paris Agreement. For cities in the Global South, there is the double challenge of climate-resilient development, that is, meeting increasing demand for housing, energy, and infrastructure for burgeoning populations, while confronting the simultaneous challenges of reducing greenhouse gas (GHG) emissions and adapting to a changing climate (UNEP & UN-Habitat, 2021). In all geographies, the implementation of transformative mitigation and adaptation in cities can be an instrument to generate livelihoods for those with lower purchasing power and enhance capacity to better respond to shocks like future pandemics, energy supply chain spasms, and food security emergencies (UNDP, 2022).

Benchmarked Learning

The *Third Assessment Report on Climate Change and Cities* (ARC3.3) builds upon preceding UCCRN *Assessment Reports on Climate Change and Cities* (ARC3), ARC3.1 (2011) and ARC3.2 (2018). The purpose of the ARC3 Series is to provide the benchmarked knowledge base for cities as they affirm their essential responsibility as climate change leaders. The ARC3 Series, with newly added ARC3.3 Elements, presents knowledge that builds on accumulated, shared experiences and thus advances and deepens with time.

In ARC3.1, cities were identified as key actors – "first responders" – in rising to the challenges posed by climate change (Rosenzweig et al., 2011). According to ARC3.1, "Cities around the world are highly vulnerable to climate change but have great potential to lead on both adaptation and mitigation efforts" (Rosenzweig et al., 2011, xxii).

In ARC3.2, this focus advanced into understanding how cities can achieve their potential by establishing a multifaceted pathway to transformation (Rosenzweig et al., 2018). It provided a road map for cities to fulfill their leadership potential in responding to climate change. According to ARC3.2, "As cities mitigate the causes of climate change and adapt to new climate conditions, profound changes will be required in urban energy, transportation, water use, land use, ecosystems, growth patterns, consumption, and lifestyles" (Rosenzweig et al., 2018, 2).

Now, as the urgency of climate change is brought home every day, ARC3.3 offers the knowledge needed to *speed up and scale up* urban action on climate change. To accomplish this, ARC3.3 presents practical methods and case study examples for accelerating change into rapid transformation in cities.

UCCRN Assessment Process

The ARC3.3 authors either self-nominated or were nominated by a third party and selected by the ARC3.3 editorial board through comprehensive vetting that prioritizes expertise, diversity, gender, and geographic balance. Each author team develops a robust assessment of an Element topic, using the latest literature, while also conducting new research. All author teams are responsible for conducting a stakeholder engagement session during the writing period, with the goal of ensuring relevance to a diverse group of urban decision-makers. During self-coined "stakeholder soundings," authors present emerging major findings and key messages to stakeholders, including city leaders from the authors' home cities, for their feedback. The UCCRN also coordinates a rigorous iterative peer-review process for each ARC3.3 Element that engages with both academic and practitioner experts, both in and out of the network.

The UCCRN's Case Study Docking Station (CSDS) is a searchable database designed to facilitate peer-learning between and among cities, benchmark actions over time, and enable cross comparisons of city case studies. The CSDS includes over 230 expert-reviewed case studies covering a range of topics such as climate change vulnerability, hazards and impacts, and mitigation and adaptation actions for sector-specific themes. The CSDS has a total of sixteen searchable variables.[1] For example, users can filter searches by climate zone, city population size, human development index, gross national income, and mitigation and/or adaptation or directly type in keywords and city names. Of the 230 total case studies, ARC3.3 authors have added 115 new ones, sharing insights, for example, on flood adaptation in Bridgetown, cloudburst planning in Copenhagen, and climate action financing in Durban.

Cities are vanguard sites for opportunities to enhance equity and inclusion, which permeate ARC3.3 in every Element as city experts delve into the multiple dimensions of climate change justice: distributive (relating to differential vulnerability of groups and neighborhoods), contextual (relating to the root causes of vulnerability), and procedural (relating to participation in decision-making for climate change interventions) (Foster et al., 2019). Elucidating ways to achieve climate justice for the most vulnerable urban groups and equal access to financial and technological resources for all cities underpins ARC3.3.

ARC3.3 Elements

The UCCRN has conducted city-centered assessments since its founding in 2007. With over 2,000 scholars and experts from cities around the world, UCCRN is addressing the research agenda that was formulated at the Intergovernmental Panel on Climate Change (IPCC) and Cities Edmonton Conference (Prieur-Richard et al., 2018).[2] Key components of this research agenda include urban planning and design; green and blue infrastructure; equity, health, sustainable production and consumption; and

---

[1] The searchable variables available to users on the UCCRN CSDS are as follows: ARC3 Assessment, Language, Continent, Country, City, Urban Design Climate Workshop, Köppen Climate Zone, Coastal (marine or riverine), City Population Classification, Urban Density, GNI Classification, HDI Index, Gini Index Coefficient, Type of Climate Intervention, Keywords. https://uccrn.ei.columbia.edu/case-studies.

[2] UCCRN actively participates in conferences that highlight the role of cities in climate change such as Innovate4Cities, the World Urban Forum, and Adaptation Futures.

finance. Over 300 UCCRN authors have now advanced this research agenda and other critical topics through the *Third Assessment Report on Climate Change and Cities* which consists of twelve peer-reviewed monographs to be published as Cambridge University Press Elements, both separately and together, throughout 2025 and 2026. This section provides a brief summary of each of the Elements.[3]

1. Learning from COVID-19 for Climate-Ready Urban Transformation

The COVID-19 pandemic has revealed gaps in city readiness for simultaneous responses to pandemics and climate change, particularly in the Global South. However, these concurrent challenges present opportunities to reformulate current urbanization patterns, economies, and the dynamics they enable. This Element focuses on understanding how COVID-19-related city-level responses impacted climate change mitigation and adaptation actions, and vice versa in terms of warnings, lessons learned, and calls to action.

2. Justice for Resilient Development in Climate-Stressed Cities

To ensure climate-resilient urban development, climate responses – both adaptation and mitigation – must include the broader city context related to equity, informality, and justice. Responses to climatic events are conditioned by the informality of the existing social fabric, institutions, and activities, and by the inequitable distribution of impacts, decision-making, and outcomes. This Element discusses differential exposure to climate events, and distributive, recognitional, procedural, and restorative justice.

3. Urban Planning, Design, and Architecture for Climate Action

Architects, urban designers, and planners are called on to bridge the domains of research and practice and evolve their agency and capacity by developing new methods and tools that are consistent across multiple spatial scales. These are required to ensure the convergence of effective outcomes across cities, regions, state/provinces, and global scales. This Element evaluates how the fields of architecture, landscape architecture, urban planning, and urban design integrate climate change mitigation and adaptation and presents a manifesto for urban transformation using science-informed design and planning.

---

[3] See www.cambridge.org/core/publications/elements/elements-in-climate-change-and-cities for the full set of ARC3.3 Elements and authors.

4. Urban Climate Science: Knowledge Base for City Risk Assessments and Resilience

Cities alter the climate system both within their boundaries and nearby through interactions with impervious land surfaces, energy generation, and transportation systems. These processes that occur on urban scales are interacting with larger-scale climate change processes to exacerbate extreme events that impact urban dwellers. This Element provides temperature, precipitation, and sea-level rise observations and projections for the cities engaged in ARC3.3 and assesses the latest research on urban heat and precipitation islands, compound extreme events, and indicators and monitoring, including the use of remote sensing in urban settings.

5. Governance, Enabling Policy Environments, and Just Transitions

The nature of governance, as a concatenation of social institutions and practices embedded at different scales, suggests the need for multilevel governance (MLG) to address the complex challenges of climate change in cities. This Element sets forth governance structures for climate action across urban, provincial, national, and international levels, analyzes the urban focus of nationally determined contributions (NDCs), and assesses the potential for urban transitions and transformations.

6. Financing Climate Action

This Element documents the availability of, and access to, finance for mitigation and adaptation in urban areas. It evaluates the current international flows, national policies, and municipal utilization capacities across private and public sectors, and nongovernmental and community-based organizations. Global financial capital is abundant but often flows to corporate investments and real estate development rather than into critical efforts to mitigate and adapt to climate change in cities. Political will and public pressure are crucial to effectively redirect these funds.

7. Infrastructure for a Net Zero and Resilient Future for Cities

Without infrastructure, cities could not exist. Infrastructure determines urban form, functions, economic development, people's livelihoods, and well-being. Using transformative infrastructure, cities can achieve ambitious GHG emission reductions, build resilience to climate impacts, and ensure inclusive and diverse access to services. This Element explores infrastructure planning concepts like circularity, decentralization, and integration and emphasizes the need for equitable, resilient systems designed according to future climate projections.

8. Nature-Based Solutions: Enhancing Capacity to Respond to Shocks and Stresses

There is growing acknowledgment that a disproportionate amount of attention and finance is invested in hard infrastructure to mitigate and adapt to climate change in cities. In contrast, soft infrastructure, that is, the use of natural features and processes, has been comparatively overlooked until recently. This Element assesses the ways that nature-based solutions (NbS) – such as reforestation, urban parks, street trees, sustainable urban drainage systems, and community gardens – can enhance the capacity of cities to reduce GHG emissions and enhance resilience to climate stresses.

9. Circular Economies for Cities

Circularity has the potential to transform cities and city systems in both the Global North and the Global South. Sustainable consumption and production, life cycle analysis, and supply and demand factors are increasingly coming into focus in cities. This Element discusses the linkages of circular economics to climate action planning, the water–energy–food system nexus, and just, local development.

10. Data and the Role of Technology

Over the past decade, changes in internet penetration and the development of new information and communication technologies (ICTs) have catalyzed an ecosystem of approaches that employ "big data" and "smart tools" to support adaptation and mitigation. Artificial intelligence and machine learning play a large role in this new technological ecosystem. This Element evaluates the opportunities and challenges for cities as they employ these new technologies and assesses emerging tools for their utility in climate change responses.

11. Perception, Communication, and Behavior

This Element explores the latest research on how urban residents perceive climate change so that the effectiveness of actions can be improved. An important corollary to this is the role that communication plays in how mitigation and adaptation actions are adopted by cities. In the event of a climate disaster, the way that cities communicate has a direct effect on residents' perception of risks and subsequent behaviors such as evacuation or strategic relocation. The Element addresses how behaviors by urban inhabitants to change mobility patterns and energy use can be encouraged in order to reduce GHG emissions, while simultaneously helping citizens to prepare for increasing climate extremes.

12. Health and Well-Being

Climate change, especially increasing extreme events, is exacerbating the risks of mortality, disease, injury, and impacts on physical and mental health and well-being in many cities. Climate change also has indirect impacts on health through disruptions in food supply and water availability. This Element assesses the latest findings on all aspects of the intersection of health and climate change for urban residents – including how urban built form, such as the presence of natural spaces, influences health and well-being under changing climate conditions.

ARC3.3 Major Findings and Key Messages

Besides the basic assessment content, each Element includes a statement of Major Findings and Key Messages. The Major Findings include statements of significant new knowledge that emerged through the assessment process, while Key Messages include recommendations for new efforts and activities with a specific focus on opportunities to speed up and scale up urban climate action.

Cross-Cutting Themes

Cities are complex social–ecological–technological systems. While ARC3.3 is composed of twelve separate Elements, together they comprise multiple synergies, interdependencies, and points of intersection. To address these connections, each Element addresses its own selection of relevant cross-cutting themes (CCTs). Figure 1 illustrates how significant recurring themes appear within the Element and the interlinkages to related Elements. Cross-cutting themes encompass such processes as drivers of urban function, change, and management; governance of cities across municipal, state/provincial, national, and international levels; and the role of city-level models and data.

Because the fundamental contribution of the ARC3 Series is to enable a learning process for urban climate action, CCTs across the ARC3.3 Elements aim to shed light on cause-and-effect relationships and elucidate

8 Climate Change and Cities

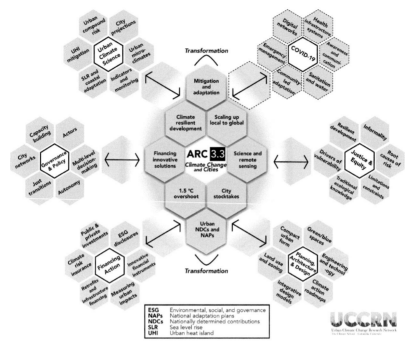

Figure 1 Cross-cutting themes associated with the overall ARC3.3 assessment and the first six Elements.

effective entry points for interventions. This focused knowledge of urban social–ecological–technological systems can inform planners, implementers, and many other city actors as they undertake ways to translate the latest science into action in their own urban communities.

Conclusion

We are pleased to launch the UCCRN *Third Assessment Report on Climate Change and Cities* with this important Element *Learning from COVID-19 for Climate-Ready Urban Transformation*. It is a prime example of UCCRN's dedication to an ongoing city assessment process. By providing both timely and benchmarked knowledge for cities as they grow into their essential role in climate change solutions, UCCRN comprehensively presents what cities need to know in order to fulfill their leadership potential. This knowledge builds on accumulated and shared experience and thus advances and deepens with time. The ARC3 Series enables this active learning process in cities all over the world, now sorely needed so that cities can indeed *scale up and speed up* their climate transformations.

Foreword I – Dr. Daniel Soranz, Municipal Health Secretary of Rio de Janeiro, Brazil

The COVID-19 pandemic was the greatest health-related challenge of the century and confronted not only health systems, but all city systems and people's ways of life globally. This Element elucidates how the COVID-19 pandemic catalyzed action, revealed inequities, and generated crucial lessons for how cities can prepare themselves in the face of climate change and future health emergencies.

This complex agenda impacts not only the health sector; this challenging period in the history of humanity highlighted deep interconnections between health, urbanization, economy, social inequalities, and the environment, bringing to light the vulnerability of cities in this confrontation. It revealed important disparities in addressing and mitigating the crisis where impacts were felt disproportionately – whether on people's health, on the population's life dynamics, or on the environment.

Preventing and confronting both health and climate crises – understanding the connection and impetus that one can give to the other – also means addressing sustainable social development to strengthen urban resilience. The economic and financial consequences, the structuring of cities, and the forms of governance for decision-making in the pandemic exposed the fragility of unprepared urban systems.

On the other hand, drastic changes in this period also culminated in experiences of innovation in city management, leading to adaptations in ways of life and mobility with reduced human activity and related transport that resulted in temporary improvements in air quality and a reduction in carbon emissions. This transformation offers an imperative to reimagine greener cities, with more sustainable transport systems and more accessible and resilient urban spaces.

Finally, this Element invites us not only to reflect, but also to act. The COVID-19 pandemic was a testimony to human resilience and the capacity for adaptation and mitigation that will vary according to the level of preparation and the resources to do so. The lessons are an invitation to rethink these connections, structures, urban resources, and an expanded vision to prepare more resilient, healthy, equitable, and sustainable cities to face climate change and future health emergencies, with the lowest possible impact on health, humanity, and the planet.

Foreword II – Lykke Leonardsen, Programme Director, Resilient and Sustainable City Solutions at the City of Copenhagen, Denmark

The COVID-19 pandemic had a huge impact on cities all around the world – including Copenhagen – although Denmark was able to manage COVID-19 more easily than many other countries.

But during COVID-19, the city – or rather the way citizens used the city landscape – changed. Suddenly, cafés, restaurants, and shops were "no go" areas, and shopping streets and centers became deserted even in broad daylight. Instead, we found that citizens were aiming for green areas in the city, along the harbor front, in parks, and all areas where it was possible to meet. Sometimes, residents gathered in such huge numbers that measures had to be taken to keep social distancing.

So, what does COVID-19 have to do with climate change? Climate change requires a paradigm shift in the way we plan our cities – and the way we use them. The COVID-19 pandemic has shown that we *can* change our behavior rapidly – and use our cities in a different way. We can take this lesson forward as we plan for the cities of tomorrow. We need to work with nature – to be able, for instance, to manage and store stormwater and to help cool cities during long heatwaves.

This can be a double win for cities. It will not only help the city to become more resilient to future climate change, but can also benefit the citizens who will have the advantage, in their everyday lives, of a greener and healthier city. This approach can also be used to upgrade neighborhoods and create better living conditions for vulnerable groups in the population.

The climate adaptation work of Copenhagen is one way of moving toward that goal. We hope that it can serve as an inspiration for other cities!

Series Editors' Introduction to *Learning from COVID-19 for Climate-Ready Urban Transformation*

Learning from COVID-19 for Climate-Ready Urban Transformation is the first of twelve publications in the ARC3.3 Cambridge University Press Elements series.[‡] It presents how cities have faced and are learning from the intertwined challenges of climate and COVID-19, with a focus on the years 2020–2022 and with special attention on the changing social and economic realities that have been presented by the pandemic shock. The authors shine a spotlight on the key finding that the COVID-19 pandemic was urban in character; 95 percent of the total reported cases were in cities and their extended metropolitan regions. They found that the varying capacities of cities in handling public health emergencies was similar to their varying capacities to meet the challenges of climate change. They document that the financial and economic impacts of the pandemic have limited cities' ability to invest in climate change solutions.

The Element assesses the global health inequities that arose, revealing the particular vulnerability of the urban poor and disadvantaged. *Learning from COVID-19 for Climate-Ready Urban Transformation* assesses the extent to which significant underlying urban issues may limit collective response capacity to confront multihazard occurrences, such as co-occurrence of the COVID-19 pandemic (or other pandemics) and increasing climate extremes in the future. It analyzes how such risks are perceived – or misperceived – particularly when urban societies confront crises that generate concatenated vulnerabilities as experienced during the pandemic. On the one hand, COVID-19 has caused unsettling disruptions to progress, and on the other, the pandemic has reinforced and interconnected human health as part of the effort to comprehensively address the dual challenges of urban health crises (i.e., COVID-19) and climate.

Since climate change is an urban threat multiplier that exacerbates human insecurities, the Element authors suggest that more coordinated urban governance is needed to enable cities to successfully respond to health and climate challenges. They document how some cities are now exploring how to integrate urban climate–environmental actions with those related to disease risk prevention and management, taking a more comprehensive urban resilience approach that embraces human health as well as climate. The overall key message from the Element is that COVID-19 has reignited the need to rethink city planning

---

[‡] Suggested Citation: Mahadevia, D., G.C. Delgado Ramos, J. Barnes, J. Fitzgerald, M. Kamei, K. Lanza. 2025. Learning from COVID-19 for Climate-Ready Urban Transformation. In Solecki, W., M. Pathak, M. Barata, A. Barau, M. Dombrov, and C. Rosenzweig (Eds.), *Climate Change and Cities: Third Assessment Report of the Urban Climate Change Research Network*. Cambridge: Cambridge University Press.

with regard to both future pandemics and increasing climate extremes, emphasizing the need for self-sustaining, walkable cities with reliable public transport, especially for rapidly urbanizing areas of the Global South.

Major Findings and Key Messages

Major Findings

1. **COVID-19 amplified disparities in health outcomes and healthcare access in cities in both the Global North and the Global South, which translated into higher mortality rates in the most vulnerable places and communities.** Nearly seven million people worldwide have died from the COVID-19 pandemic, as of October 2023. These health risks are concurrent with climate change risks experienced by low-income and disadvantaged communities. The urban poor and those in the low-end informal sector are the most vulnerable populations in cities, particularly women with mobility constraints (Section 2).

2. **From 2020 to 2022, the COVID-19 pandemic was urban in nature, with 95 percent of total reported cases in cities.** Given rapidly accelerating world urbanization (55 percent of the world's population now live in cities with highest growth rates in South Asia and sub-Saharan Africa) and intensifying climate change induced extreme events occurring in many cities, it is critical to understand the interlinkages between urbanization, climate change, and COVID-19 (Section 3).

3. **Climate change-related disasters were coincident with the COVID-19 pandemic in cities.** Cyclone Amphan in Eastern India resulted in eighty cyclone-related deaths, mass evacuation and sheltering, and an associated spike in COVID-19 cases. In several US cities, public cooling centers had to close in the summer of 2020 because their cooling systems did not have the mandatory air filter to avoid COVID-19 propagation, ultimately increasing the risk of morbidity and mortality due to both heat waves and COVID-19 (Section 3.2).

4. **Evidence shows that urban population density did not play a major role in COVID-19 propagation in cities, even though density has often been considered a key indicator of social vulnerability to disasters.** On the contrary, urban spatial form and organization have been shown to be more relevant than population density in COVID-19 transmission, given the statistics that higher-density cities had lower infection rates. Higher-density coupled with effective urban design (mixed-used, mid-rise residential buildings within an urban fabric featuring wider sidewalks, inclusive and safe green and public spaces, and more cycling lanes) are urban features that

support the safety and health of urban populations during both pandemic and changing climate conditions (Section 3.1).

5. **After a brief period of reduction in GHG emissions at the beginning of the pandemic, many cities experienced increased dependence on fossil-fuel based cars for personal mobility, raising emissions of GHGs.** This was evident in cities in India, Australia, Canada, USA, and Germany, in direct opposition to climate change mitigation efforts promoting a shift to public transport. Alternatively, some cities promoted bicycle transportation, especially in Europe, where cycling rates increased from 15 to 27 percent from 2019 to 2021. However, resistance to changes in modes of personal mobility during the COVID-19 emergency has been identified in some car-dominant cities, such as Monterrey City, Mexico, despite efforts to encourage active mobility (Section 4.2).

6. **Globally, renewable energy usage reached an all-time high during the pandemic, indicating that renewable energy resources have the potential to surpass traditional fossil fuels.** There is a need to refrain from returning to pre-COVID-19 business-as-usual practices and, instead, transition to sustainable, renewable energy use that proved to be efficient during the pandemic. Utilization of renewable energy sources, such as wind and solar, and advancing financing of these projects in cities is necessary to facilitate a zero-carbon energy transition (Section 7.1).

7. **The economic and financial impacts of the COVID-19 crisis limited cities' ability to invest in climate change adaptation and mitigation efforts such as clean energy, public transport, and the creation of blue and green infrastructure.** Some efforts have linked pandemic recovery to sustainable, climate-resilient, and just and inclusive investments. However, the results of these efforts have been mixed and were affected by slow economic recovery and inflation (Section 7).

8. **Some countries experienced a centralization of decision-making and autocratic impetus during the pandemic, making cities dependent on national governments for rules, protocols, equipment, financing, and human resources; other countries and cities formed multiscale partnerships.** In some cases, COVID-19 enabled grassroots movements to pressure cities to expand green areas and public spaces, integrating active mobility pathways. These experiences are important for planning effective governance of combined health and climate crises in the future (Section 6.1).

Key Messages

1. **Urban responses to COVID-19 offer hope for cities' ability to tackle the long-term challenge of climate change.** Well-established intra- and intercity partnerships can support future climate and health emergency response as well as climate change mitigation and adaptation planning. The multiscalar and partnership-based efforts developed to address the COVID-19 challenge, if institutionalized, provide opportunities for additional long-term coordinated climate change mitigation and adaptation efforts.
2. **Health and climate solutions in cities should be cogenerated and implemented with continuous consideration and active participation of the most vulnerable groups including low-income people and informal dwellers.** Low-income populations, particularly ethnic groups, tend to live in "structurally vulnerable neighborhoods" that are usually disregarded and neglected.
3. **COVID-19 revealed that cities, particularly low-income, rapidly urbanizing cities in Asia and Africa, need to address health, socioeconomic, cultural, and governance issues to enhance their capacity to respond to health and climate crises.** Multilevel governance and local management strategies can help reduce risks from present and future climate extremes in both the Global South and the Global North.
4. **Support for digitalization and associated technologies can improve pandemic responses and climate change management practices in the future.** Countries and cities with higher technology adoption and digitalization generally showed a greater capacity to adapt to the COVID-19 pandemic and implement institutional interventions.
5. **Infrastructure, land use, and housing interventions can provide synergies between COVID-19 and climate mitigation and adaptation responses.** Examples include repurposing streets as public spaces or for advancing active mobility, adapting public transport operations, maintaining nonpharmaceutical advisories, and setting up recovery facilities in dense informal housing areas.

1 Introduction and Framing

At the beginning of 2020, cities all over the world were simultaneously managing and adapting to the SARS-CoV-2 virus, widely known as COVID-19,[4] which caused a devastating global health emergency. While the ARC3.3 COVID-19 Element focuses on the public health emergency caused by COVID-19, the assessment sets a precedent for evaluating the implications of other major events at global, national, or city levels – such as other epidemics, conflicts, and wars, and even economic collapse – on actions related to climate change. Any major disruption or emergency has the potential to divert government and nongovernment organizations to meet the immediate emergency rather than paying attention to the long-term actions needed to address climate change adaptation and mitigation. This Element investigates the interactions of COVID-19 and cities and assesses COVID-19's implications for urban climate change action and the most vulnerable communities.

Going forward, even as incidences of COVID-19 continue to wax and wane across the world, cities will need to address the challenges of climate change, public health emergencies (including epidemics and pandemics), poverty, and inequity effectively and simultaneously. Cities must overcome obstacles to develop more robust, inclusive, equitable, healthy, sustainable, and resilient pathways – a process that requires an understanding of globally interconnected cities and complex urban systems, and furthermore, of global, regional, and local socioeconomic, cultural, political, and innovation dynamics (UNEP & UN-Habitat, 2021). The COVID-19 pandemic has presented opportunities to reformulate current urbanization patterns, urban economies, and the dynamics that cities enable (UNEP & UN-Habitat, 2021).

The COVID-19 public health emergency has provided an opportunity to assess the linkages among Sustainable Development Goals (SDGs) such as climate change actions (SDG 13) and actions related to health (SDG3) within the framework of resilient and inclusive cities (SDG11). Hence, this Element is a cross-cutting one, analyzing both the synergies and cobenefits of COVID-19 actions and climate change responses at the city level as well as the impacts and disruptions. The aim is to provide insights for future climate change actions and pandemic (and other emergency) preparations. The framework of synergies and cobenefits and trade-offs utilized here is based on the IPCC's 1.5°C report (IPCC, 2018) and Sixth Assessment Report (AR6) (IPCC, 2022d).

This widespread health crisis – addressed with a range of health infrastructure and access measures – has had differentiated disruptive implications on social,

---
[4] The COVID-19 pandemic was declared by the World Health Organization on March 11, 2020 due to the rapid global spread of the disease, mostly in urban areas. It was the first major global pandemic since the 1918–1919 influenza pandemic (WHO, 2020c; Hu et al., 2021).

economic, and political dynamics (Pelling et al., 2021). The pandemic led governments, both national and municipal, to divert financial resources toward immediate healthcare needs and the development of vaccinations. The COVID-19 pandemic has revealed gaps in city readiness for pandemic response, particularly in the Global South (Delgado Ramos & López, 2020; Sverdlik & Walnycki, 2021), and in climate change adaptation and mitigation in cities (Dodman et al., 2022; Lwasa & Seto, 2022). There are documented setbacks in educational achievements, reductions in life expectancy at birth, and in public and private investments in climate mitigation technologies (Hourcade et al., 2021; Londsdale et al., 2020; UNDP, 2022).

Climate change actions as well as COVID-19 – or any other pandemic – response and preparedness interact with physical and nonphysical urban features. Figure 2 frames the complex and multiscale nature of such interactions, both hazard and nonhazard dependent, through the lens of urban exposure, sensitivity, and vulnerability to shocks and their associated risks and impacts. It also presents the interaction of climate and health actions in

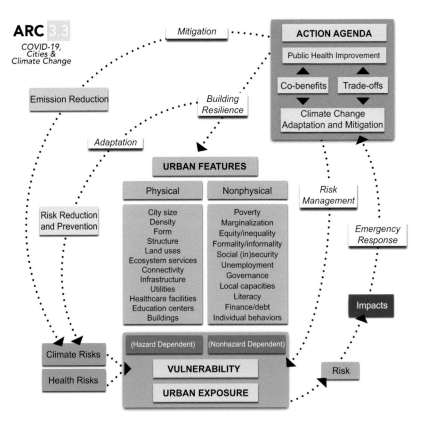

Figure 2 COVID-19 and climate change interactions at the urban scale.

cities through the potential synergies and expected trade-offs between adaptation, mitigation, and resilience building. The identification of both synergies and trade-offs can help to calibrate, align, and integrate actions that may successfully "work for people and planet" by preventing, reducing, and managing urban risk in complex and multihazard scenarios (UNEP, 2021).

The multidimensional spatial and temporal implications of the COVID-19 pandemic – including those derived from lockdown and other nonpharmaceutical interventions – are assessed by analyzing their relationship with:

- Urban **climate change related hazards** like compound risks (i.e., those occurring simultaneously) through cascading impacts (i.e., those leading to effects in multiple sectors sequentially) are unevenly experienced by cities and their inhabitants;
- Urban **planning and design**, including physical urban features such as density, urban form, sprawl, and uses and connectivity;
- Urban **built infrastructure**, including physical urban features related to buildings, transportation, energy, and water resources;
- Urban **ecology**, including physical urban features associated with nature-based solutions and ecosystem services;
- **Waste generation and management**, including physical and nonphysical measures for advancing urban circularity;
- **Consumption behavior changes and urban supply chains**, including physical and nonphysical urban features related to urban logistics, resource efficiency, and circularity;
- Urban **informality, poverty, and inequity** as nonphysical conditions that aggravate social vulnerability and reduce urban resilience.

The diverse components of this framing are connected to other ARC3.3 Elements as well (Figure 3). For example, during COVID-19, governance and emergency responses were critical to pandemic adaptation and community well-being. The pandemic also brought forth questions on how physical aspects of cities, such as size, density, form, structure, and land use can affect the spread of disease. Additionally, in the presence of compound climate and health events, rapid financial relief can influence the extent to which a city is able to cope with simultaneous issues that require immediate response. Lastly, equitable access to healthcare within and across cities can reduce human vulnerability to health and climate emergencies.

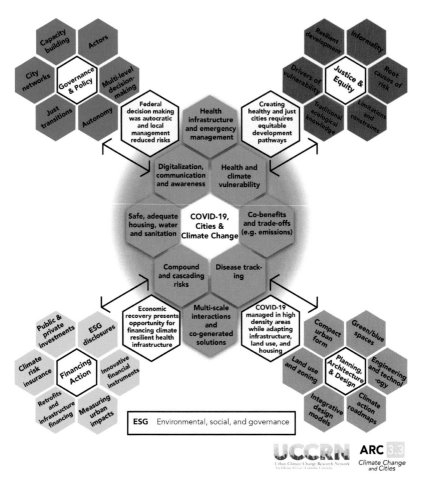

Figure 3 Cross-cutting themes linking ARC3.3 COVID-19 to other Elements.

Learning from COVID-19 for Climate-Ready Urban Transformation is organized into eight sections.

- **Section 1, Introduction and Framing**, provides background information and frames COVID-19 and climate change in cities as well as introducing themes and CCTs from upcoming ARC3.3 Elements.
- **Section 2, Vulnerability and Informality**, analyzes the unequal impacts of COVID-19 and climate on vulnerable and informal urban populations.
- **Section 3, Interactions at the Urban Scale**, explains how urban characteristics influence COVID-19, compound risks, and air quality, using the latest data.
- **Section 4, Urban Systems: Built Environment, Transportation, and Waste**, describes the complex interactions between COVID-19 and the built environment as well as the transportation, waste, and water sectors.

- **Section 5, Links with Urban Ecology**, investigates the relationship between access to green spaces and COVID-19 case rates.
- **Section 6, Governance and Urban Climate Action**, evaluates emergency decision-making, multilevel governance, leadership, social partnerships, and technology. It also identifies enablers and barriers in advancing transformative agendas.
- **Section 7, Energy and Economics**, assesses nonrenewable and renewable energy consumption, supply chain disruptions, and economic impacts.
- **Section 8, Learning from COVID-19: Accelerating Urban Transformational Pathways**, considers the lessons learned and opportunities for building pandemic-resilient cities through the advancement of low-carbon, sustainable, inclusive, and just urban transformation pathways.

2 Vulnerability and Informality

A novel coronavirus, COVID-19, was identified on January 9, 2020, and declared as a global pandemic by the World Health Organization (WHO) on March 11 of that year (Hu et al., 2021). As of October 2023, nearly seven million people worldwide had died from COVID-19, and there had been almost 775 million confirmed cases (WHO, 2023). The COVID-19 pandemic has amplified disparities in health outcomes and healthcare access and this has translated into higher mortality rates in the most disadvantaged places and communities (McGowan & Bambra, 2022; WHO, 2020a).

Because climate change and COVID-19 have unequal effects on populations – given social, economic, and political inequalities, inequitable pathways of development, and inactions to address climate change thus far – cities need to account for physical and nonphysical urban features, mitigate future risk, and seek cobenefits and trade-offs with climate action to create climate- and pandemic-resilient cities (Cities Alliance, 2021). The varied success of cities' responses to COVID-19, however, has underscored how poverty, inequity, and informality exacerbate urban vulnerability, compromising long-term sustainability, resilience, and community well-being. The economic and financial consequences of COVID-19 have shifted underlying vulnerability in many ways, and this requires attention in both development and climate change responses.

2.1 Vulnerability

The COVID-19 pandemic has uncovered many underlying socioeconomic inequities and uneven urban living conditions that have been exacerbated by the pandemic and, in some cases, by the measures taken to cope with the COVID-19 emergency or due to a limited local capacity to act (Boza-Kiss et al., 2021; Delgado Ramos & López, 2020). The *Sixth Assessment Report of the United Nations Intergovernmental Panel on Climate Change* (IPCC AR6) flags the issue of the higher vulnerability of low-income households and regions to the adverse impacts of climate change (Dodman et al., 2022). Low-income populations, particularly minority ethnic groups, tend to live in "structurally vulnerable neighborhoods" that are usually disregarded and neglected (Berkowitz et al., 2020; Delgado Ramos et al., 2020).

In São Paulo, COVID-19 patients living in the bottom 40 percent of the poorest areas were more likely to die when compared with patients living in the top 5 percent of the wealthiest areas, indicating that low-income patients had a higher mortality risk than high-income patients (Li et al., 2021). Another study, from Huzhou, China, reported the following groups as vulnerable to COVID-19 infections and the adverse impacts of lockdowns: lower-income individuals; small communities with insufficient green spaces; tourism, transportation, and catering workers; those working in and owning informal businesses; and students (Yang et al., 2021).

In addition, migrants and migrant workers tended to be more vulnerable since they were usually excluded from any local planning and aid during lockdowns (Bai et al., 2020). This was – and still is – particularly concerning in the case of massive migrations of poor populations – for example, migrants traveling from Central America to Mexico and to the USA. This has foregrounded the need to reformulate cross-border cooperation due to higher levels of confirmed cases and deaths when compared with national averages, which are even higher in the case of refugee camps (Arriola Vega & Coraza de los Santos, 2020; Lara-Valencia & García-Pérez, 2021). In Singapore, too, migrant workers living in crowded dormitories were more affected by infection than other population groups (McFarlane, 2023). Similarly, during climate change extreme events, migrants tend to be overlooked in relief and recovery efforts. As an outcome of COVID-19 and disaster-reduction learning, the basic needs of migrants should be included in long-term planning and preparedness.

In cities in the Global North, low-income populations were more likely to experience employment loss (ECLAC, 2020; Gil et al., 2021) or to have low-paying jobs which put them at higher risk of COVID-19, due to food and nutrition insecurity, among other reasons (Florida & Gabe, 2023; Ruszczyk et al., 2021). In cities of the Global South, COVID-19 as a multiplier of existing multidimensional vulnerability has, for instance, been most noticeable in informal settlements in India where chronic distress has grown among the urban poor (Bhide, 2021; Indorewala & Wagh, 2020). (See the additional resources for a further literature review of COVID-19 vulnerability in the Global North and the Global South.)

2.2 Informality and Managing COVID-19 in Informal Settings

The inclusion of informal settlements and poverty conditions in climate change adaptation and mitigation pathways is presently limited; key areas include the need for increased housing, especially for those transitioning from informal to formal housing (Lwasa & Seto, 2022). Lockdown measures to contain COVID-19 infections increased inequities and had severe adverse impacts on the poor, especially women (WIEGO, 2020). Stringent lockdowns in the Global South were particularly disastrous for those living and working in the informal sector (Botello Peñaloza & Guerrero Rincón, 2022; Khambule, 2022; Swarna et al., 2022). In some cases, lockdowns meant that street vendors could no longer operate (Moctezuma Mendoza, 2022; Murillo, 2022); auto-rickshaws, tuk-tuks and other informal transport modes were pushed out of the street and the drivers either lost their income, or saw it drop (Spooner & Whelligan, 2020). Service-sector workers, such as domestic helpers, were not permitted to enter their places of work (Dogar et al., 2022).

Informal waste recyclers no longer had access to spaces in the city that sustained their livelihoods and, when they did, were exposed to COVID-19 related waste (Hartmann et al., 2022; Harvey, 2022). Prices of recyclable materials also dropped, putting at risk the livelihoods of thousands of informal recyclers, particularly in South Asia where 82 percent of the recycling trade is carried out by informal waste pickers (Hicks, 2020; Nguyen, 2020; UNU-WIDER & WIEGO, 2022). Box 1

BOX 1 THE COVID-19 CRISIS AND THE INFORMAL ECONOMY

Women in Informal Employment: Globalizing and Organizing (WIEGO), a UK-based charitable company, has evaluated COVID-19 impacts on informal work in a sample of cities. Two survey studies were carried out during and after peak lockdown (WIEGO, 2020, 2022). The findings are:

- By mid 2021, informal workers' earnings had risen by 64 percent of their pre-COVID-19 level – almost doubled from mid 2020.
- Home-based workers were particularly impacted by the pandemic. "Live-out" domestic workers' earnings had recovered to 91 percent of average pre-COVID-19 earnings by mid 2021, yet with large differences among cities.
- Informal street vendors struggled to recover, particularly informal women vendors. By mid 2021, 90 percent of vendors were back at work, but their earnings were only 60 percent of pre-COVID-19 levels.
- Waste pickers faced fluctuating prices and difficulties in accessing waste.

- To cope, informal workers' and their families used their savings (in 35 percent of cases), borrowed money (46 percent), sold assets (9 percent), or pawned assets (5 percent).

The WIEGO studies show major challenges for people in the informal sector, including difficulties finding work and maintaining pre-COVID-19 earnings (WIEGO, 2022). The implication, despite the existence of variable government cash relief, was that COVID-19 exacerbated food insecurity and hunger, especially in Lima in South America and in South Asian and African cities. The data on the high vulnerability of informal workers during the COVID-19 pandemic is clear, and it supports the known fact that low-income households and those employed in the informal sector are more adversely affected by disasters, especially climate disasters (Revi et al., 2014).

shows examples of the adverse impacts of COVID-19 on self-employed people in selected South Asian and African cities.

Understanding the short- to long-term implications of actions taken by the most vulnerable urban groups is of utmost importance for identifying the challenges, trade-offs, and potential solutions for taking forward inclusive, just, and sustainable recovery plans. Working with residents to cogenerate appropriate solutions, particularly in poor and informal urban areas, is of great importance when facing not only a pandemic crisis but other hazard situations as well. Addressing the degree to which urban areas and their inhabitants, which carried most of the COVID-19 burden, overlapped with those affected by climate change impacts and environmental degradation may inform where work should be prioritized to enable a more robust but also integrated approach to building long-term alternative-solution pathways (Bununu & Bello, 2024).

Moving toward alternative governmental pathways implies collaborating with nonstate organizations and resident-led/grassroots initiatives. Residents can offer valuable first-hand accounts and coproduce solutions that fit into specific configurations of formal and informal settlements (Fahlberg et al., 2020, 2023). Resident-led initiatives responded to basic local needs during the COVID-19 pandemic through networked participatory processes, from those related to water provision (including hand-washing stations like those installed in Lesotho, Ghana, Burkina Faso, Namibia, Zimbabwe, and Kenya) and waste collection, to food accessibility schemes that work with short food-supply chains, the philanthropic provision of food, personal protective equipment (PPE) for clinical staff, city residents schemes, and even data monitoring related to disease incidence and response at the local level (Achremowicz & Kamińska-Sztark, 2020; Cortis & Blaxland, 2020; Loewenson et al., 2020; Tageo et al., 2021; UN Women, 2021).

Similar actions can be tracked for social participation, coproduction of knowledge, and cogeneration of climate change solutions and tracking practices at the urban scale, promoting partnerships and alliances that have been emphasized as particularly relevant for accelerating urban transformation (Solecki et al., 2021; WCRP, 2019). These experiences and, more generally, a broader effort in exploring how democratic science policy–practice may take place at different scales, can accelerate social participation and capacity building to support urban transformation processes.

Megacities in developing countries, which typically have large populations in informal settlements, have few traditional welfare provisions but high potential for knowledge coproduction. These cities tend to lack demographic and socioeconomic information related to informal settlements, making it difficult to extend regular welfare measures and monitor the spread of a pandemic or the impacts of a disaster. One such megacity is Mumbai with a population of over twenty million people and Asia's largest slum, Dharavi. Using decentralization strategies and involvement from a wide range of government and nongovernment stakeholders, Dharavi successfully managed the spread of COVID-19 during a second wave from April to June 2021 (see Case Study 1). Through strategic efforts and a culture of voluntarism, the spread of COVID-19 was contained in Dharavi, and Mumbai did not experience as many deaths as Delhi and other cities in northern parts of India.

In general, community groups were central in responding to the COVID-19 emergency. Many local groups distributed food and water to people living in the most vulnerable slums, those who were homeless, and migrant workers living in temporary shelters. In Latin American cities, collectives like civil society organizations (CSOs), which were already working with communities on myriad issues such as housing, income generation, food security, human security, public health, infrastructure, and political participation stepped up efforts to manage COVID-19. Community-based organizations and their partnerships with government and nongovernment actors will be essential in managing climate change and pandemic-induced disasters and emergencies. (See the additional resources for further information on local COVID-19 efforts.)

CASE STUDY 1 COPING WITH THE PANDEMIC: THE DHARAVI MODEL[5]

Amita Bhide

COVID-19 testing in Dharavi (Source: Karan Vijay Sharma, 2020).

Dharavi is one of the world's largest, most densely populated informal settlements, with about one million people in a space of 2.1 square kilometers. Housing is largely comprised of kutcha (nonpermanent) houses, ranging from 10–15 square feet in dimension and often occupied by more than ten people per unit. Although about 60 percent of residents have access to shared tap water and 78 percent have access to shared toilets, nonnetworked facilities result in poor water and sanitation quality. Socioeconomic vulnerabilities deepen residents' susceptibility to crisis.

---

[5] See extended version of case study at https://uccrn.ei.columbia.edu/case-studies.

Despite a lockdown, from April to May 2020 Dharavi recorded about fifty deaths per day. During this time, the settlement was subject to a cascading crisis – the COVID-19 pandemic combined with a hunger emergency due to loss of work brought about by the lockdown. Because of this, eliciting the cooperation of Dharavi residents was challenging. However, a series of actions by the authorities in partnership with local people ensured that Dharavi was able to contain pandemic infections within a fairly short time frame. In the process, this effort outlined pathways for urban climate resilience in cities in the Global South where informality and lack of public infrastructure are key features.

The COVID-19 response was mobilized in two centers by converting hospital resources. In Community COVID Care 1, high-risk groups were isolated, and in Community COVID Care 2, people who were COVID positive were quarantined. The local doctors' association was enlisted to run these centers and undertake COVID-19 screenings. School staff and hall managers converted school assembly and wedding halls into quarantine centers. Partnerships, nongovernmental organizations (NGOs), and local entrepreneurs mobilized food aid for those in the centers and families facing hunger. Several innovations in contracting, designing, engineering, and recruitment ensured scalability and the financial viability of these facilities.

A data system of referrals was created to deliver case-specific treatment, despite limited hospital beds and trained human resources. High COVID-19 case areas were mapped. High-risk spots such as public toilets were identified for frequent sanitization. Facilities were monitored to ensure optimal utilization, and user feedback was incorporated into improvements. Psychosocial and physical care were deemed equally important, and recreational activities such as yoga, meditation, dance, and singing boosted morale.

During the second pandemic wave in March 2021, Dharavi experienced a sharp but brief spike in case rates because the necessary systems were in place and vaccinations were available. Despite adverse pandemic conditions, Dharavi has emerged as a model of resilience. As the unprecedented impacts of climate change worsen and unevenly effect informal settlements, the Dharavi model offers a path forward that is based on recognizing the urban poor as citizens, complementing their participation with decentralized and localized responses, and respecting their dignity.

2.3 Role of Informal Housing, Water, and Hygiene

Buildings are central to an overall urban resilience strategy as well as specific resilience to climate change impacts. Beyond housing density, specific housing features have been associated with the spread of COVID-19 (and the health, safety, and well-being of residents in general). For example, fresh air and natural light have a positive impact when managing a pandemic. Natural ventilation and smart solar shading can alleviate heat impacts in summer and reduce heating requirements in winter.[6] Further research is needed on the links between housing design and pandemic spread.

Some exploratory studies have been carried out in Cairo, Egypt, and found that ventilation and natural light are essential features in homes, followed by availability of outdoor space and at least one bedroom with its own bathroom for isolation needs (Alhadedy & Gabr, 2022). A similar study has been carried out on multiunit residential buildings in China that offer differentiated feature priorities by age, gender, and occupation (Xu & Juan, 2021).

The relationship between density, disease transmission, and climate resilience is complex, especially with respect to heat and informality. Low-income populations tend to live in highly dense conditions, for example in Santiago, Chile (Valenzuela-Levi et al., 2021), Lisbon (Mendes, 2020), cities in France (Goutte et al., 2020), and those in New Jersey, USA (Okoh et al., 2020). While high density is promising as a climate change mitigation effort, if it is not properly designed it may worsen heat-island effects (Dodman et al., 2022).

Significant populations in the Global South live in informal housing with roofing materials that absorb and transfer heat to interiors. Living in such conditions during a lockdown period can lead to heat stress (Mahadevia et al., 2020). Informal neighborhoods also tend to have narrow streets that do not permit breezes to flow through them (Satterthwaite et al., 2018). Thus, lack of house and street ventilation increased risk of COVID-19 spread as well as thermal discomfort (Bhide, 2021). In addition, inadequate water supply in informal settings made it difficult to maintain hygiene during the COVID-19 pandemic as well as to be able to use water for cooling.[7] Formal, affordable, and well-built housing with better ventilation offers resilience to climate extremes and, possibly, pandemics.

A major challenge is that new housing for low-income households transitioning from informality tends to be built on urban peripheries. This contributes to

---

[6] See ARC3.3 Element on *Urban Planning, Design, and Architecture*, www.cambridge.org/core/publications/elements/elements-in-climate-change-and-cities.

[7] See ARC3.3 Element on *Equity, Development, and Informality*, www.cambridge.org/core/publications/elements/elements-in-climate-change-and-cities.

urban sprawl and means larger commuting distances (Coelho et al., 2020), heightening dependence on motorized transport in the absence of poor transit connectivity. While other income groups manage mobility through private transport, low-income groups have continuing mobility challenges. More distant locations also reduce the possibility of active transport options, that is, cycling and walking, which were considered safe options during the COVID-19 pandemic (Valenzuela-Levi et al., 2021). Additionally, peripheral housing may create vertical slums that exacerbate the overall vulnerability of inhabitants and reduce their resilience to climate and pandemic impacts.

An important factor in COVID-19 resilience is permanent housing, which is not the case for individuals experiencing homelessness, who number over 150 million individuals worldwide (UN-Habitat, 2020a). Housing evictions were another specific risk factor during the pandemic (Benfer et al., 2021). Without a home, the risk of exposure to heat and other adverse weather events as well as disease susceptibility increase (Bezgrebelna et al., 2021; Kidd et al., 2021), particularly for children – the pandemic orphaned and pushed millions of children into homelessness (Unwin et al., 2022). An estimation based on excess deaths data – ignoring inconsistencies in COVID-19 testing and incomplete death reporting – projected that up to 10.5 million children lost parents or caregivers and about 7.5 million experienced COVID-19 associated orphanhood between January 1, 2022 and May 1, 2022 (Hillis et al., 2022).

As a multiplier of preexisting vulnerabilities, such as water poverty, water scarcity, and growing underserved informal urban populations, the COVID-19 pandemic has deepened and accelerated water insecurity. Capacity has been diminished to guarantee basic sanitary and hygiene conditions that are at the core of preventing disease and protecting human health (WHO, 2020c). Water and sanitation access are also important for a resilient city, either for dealing with heat extremes or maintaining hygiene during COVID-19 and other pandemics.

For the urban poor and residents of informal settlements in the cities of the Global South during the COVID-19 pandemic, the lack of secure access to water and to sanitation became a double burden. Water demand, especially household demand, increased during the COVID-19 pandemic (Buurman et al., 2022; Nemati & Tran, 2022), raising water security concerns in cities where availability was already constrained due to overexploitation and climate change impacts on freshwater sources. Utilities were unable to recover operational costs during the pandemic, either due to customer inability (as in the case of Los Angeles and another dozen large US cities [Walton et al., 2021]) or unwillingness to pay for poor service provision (as in the case of Uganda's households [Sempewo et al., 2021]), and have faced serious financial challenges that, in some cases, were already an issue before the COVID-19 pandemic. Such financial constraints may lead to blockages in essential supply chains for water, sanitation, and infrastructure maintenance.

Financial limitations and local realities have obliged local authorities to absorb the cost of improving or expanding water service capacity in a short period of time, rather than offering disinfectant gel to populations in areas that lack regular water provision. This was the challenge in areas of Mexico City experiencing conditions of water poverty, as reported by the former head of the National Water Commission (Jiménez Cisneros, 2020). In some cases, because water provision is directly linked to public health and well-being, water bills were subsidized or fully funded during lockdown by national or subnational governments in rural and urban areas; this was the case across a range of countries, from Bolivia and Thailand to Ghana (Nunoo, 2020; Serrano & Gutierrez Torres, 2020; Xinhua, 2020).

However, these are not long-term solutions, like improvement in water supply and efficiency (PNUMA, 2021). (For details of Mexico City's water harvesting program, which received the 2022 International Council for Local Environmental Initiatives (ICLEI) prize for the city government with the best water management, see "Mexico City's Rainwater Harvesting Program" on UCCRN's CSDS [Delgado Ramos, 2021].)

If correctly designed for sustainability and resilience, improving water security can reduce health and climate vulnerability and contribute to mitigating GHG emissions while supporting processes that address environmental and food security issues. For this reason, the COVID-19 pandemic can be used as an opportunity to advance schemes for the integrated management of water–energy–food resources for urban consumption (Tayal & Singh, 2021).

3 Interactions at the Urban Scale

The COVID-19 pandemic was urban in nature with 95 percent of total reported cases located in urban areas (UN-Habitat, 2020a). Its spread occurred in waves of different and newly emergent variants of the coronavirus over differing time trajectories.

3.1 Density and Size

Urban climate change mitigation often focuses on establishing higher levels of density to reduce sprawl and hence car dependency (IPCC, 2022d; UNEP & UN-Habitat, 2021).[8] Higher urban density often influences climate exposure,

---

[8] See ARC3.3 Element on *Urban Planning, Design, and Architecture*, www.cambridge.org/core/publications/elements/elements-in-climate-change-and-cities.

sensitivity, and the adaptive capacity of cities (Teller, 2021). In the initial days of the COVID-19 pandemic, prevalence of the disease was assumed to be linked to high densities and crowded living conditions. It was assumed that dense and compact cities like New York, Paris, London, Madrid, Rio de Janeiro, Mexico City, New Delhi, Mumbai, Manila, and many more would be impacted greatly by the fast-spreading impacts of COVID-19 (Teller, 2021). Density has often been considered a key indicator of social vulnerability – that is, dense areas are locations where the poor are concentrated – especially in the event of natural disasters (Fatemi et al., 2017). Hence, it was expected that this could be an important cause of COVID-19 propagation; however, the evidence suggests otherwise.

Urban densification has not consistently been related to greater transmission or mortality rates during the COVID-19 pandemic in cities in the USA (Angel & Blei, 2020) and Asia (Hong Kong, Indonesia, Singapore, and Tokyo [Patino, 2020; Wahid and Setyono, 2022]). Data from 384 US metropolitan statistical areas analyzed between March and July 2020 showed that cities with double the population density of cities with lower density were projected to have 4.1 percent fewer cases per capita and 7.4 percent fewer deaths (Angel & Blei, 2020). Another data set, collected from the USA in 2021, indicated that density was not the sole factor in the spread of COVID-19, and that this was, in particular, due to behavioral adaptations that may take place through risk-compensation processes (Paez, 2021).

A study mapping the spatial geographies of COVID-19 infections across three waves combined with heat vulnerability in New York City showed that heat and COVID-19 exposure were influenced by natural, built, sociodemographic, and environmental factors including access to green infrastructure (Knox-Hayes et al., 2023). The New York study found that COVID-19 risk geographies changed across the three waves of the pandemic, and before the onset of the third wave, due to the impact of vaccinations. In the case of twenty-seven Brazilian federal units, four months following the first mortalities an increase of 1 percent in population size (not density) was correlated with a 0.14 percent increase in the number of fatalities per capita (Ribeiro et al., 2020). A study in Ahmedabad, India, with data up to mid July 2020, observed that COVID-19 infections were not related to density but to the extent of implementation of WHO nonpharmaceutical interventions and general awareness of masking, social distancing, and sneezing or coughing into one's elbow (Mahadevia et al., 2022).

Thus, the available literature does not directly link the spread of COVID-19 infections with urban density. On the contrary, proper levels of density coupled with effective urban design (mixed-used, mid-rise residential buildings within an urban fabric featuring wider sidewalks, inclusive and safe green and public spaces, and cycling lanes) may be features that support the safety and health of urban populations during a pandemic (Marconi et. al., 2022; UNEP & UN-Habitat, 2021).

A second popular idea was that the spread of infection was faster in large cities than in smaller ones. When COVID-19 infections per 100,000 population were plotted against the populations of individual cities, the relationship between city size and infections was true for European cities, but not always for cities in the Americas, South and Southeast Asia, the Eastern Mediterranean, Western Pacific, and Africa (Figure 4). All large cities in Europe experienced a high incidence of infections while the largest cities of South and Southeast Asia – Mumbai, Dhaka, Jakarta, Karachi, and Bangkok – reported low incidences of COVID-19 infections.

In the Americas, infections per 100,000 population were low in Buenos Aires, São Paulo, and Santiago but high in Bogota, Chicago, and Montreal. More specifically, statistical analysis in the USA indicated that COVID-19 cases and fatalities were higher per capita in more populated cities (Angel & Blei, 2020). Cases in African cities were exceptionally low, which could be linked to mobility and reporting. Therefore, while it may be argued that urban spatial form and organization are more relevant than population density in COVID-19 transmission, further research is required (Blanco, 2020). These assessments, moreover, are contingent upon the quality of reported data, which may vary significantly across cities.

Larger cities, despite their relative advantage, may be comparatively more vulnerable due to interdependencies with other urban and nonurban locations. Response capacities, however, seem to be relatively correlated to city size as bigger cities tend to have more infrastructure, budgets and staff, as well as access to credit and international loans (Figure 5). The inverse seems not to be the case, as city size has not been found to be as significant as urban governance capacity when dealing with COVID-19 prevention and control (Chu et al., 2021). The varying capacities of cities to handle public health emergencies like COVID-19 may reflect differing abilities to meet climate-related challenges and, furthermore, uneven capacities for implementing coordinated agendas on climate–environmental and health issues at both local and metropolitan scales (Delgado Ramos, 2021; ECLAC, 2021).

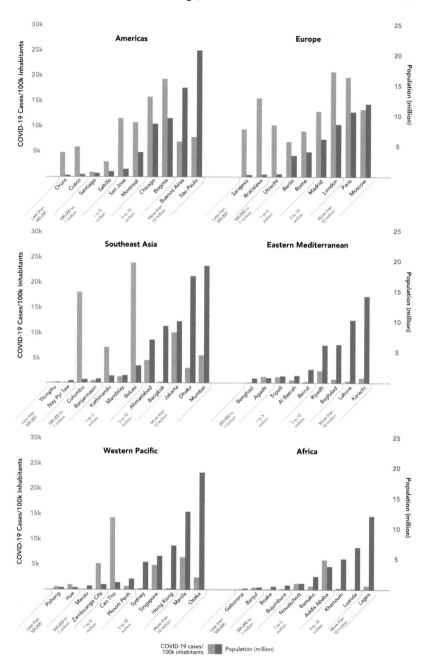

Figure 4 COVID-19 confirmed cases for the six WHO regions by city size (based on data from https://unhabitat.citiiq.com, accessed January 3, 2022).

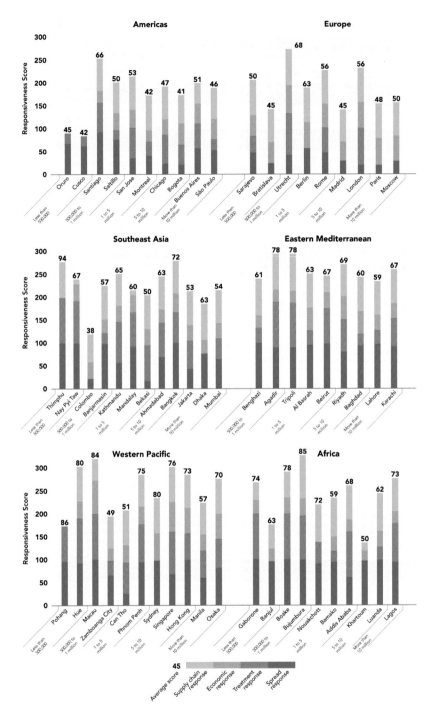

Figure 5 Degree of responsiveness in a selection of cities in the six WHO regions (2020–2022) (based on data from https://unhabitat.citiiq.com, accessed January 3, 2022). The responsiveness score is provided by UN-Habitat and is based on spread response, treatment response, economic response and supply chain response.

3.2 Compound Risks with Climate Hazards

Looking to the future, cities will have to learn to respond to the predicted increase in the number of epidemics, pandemics, and climate extremes (Marani, 2021). This presents an opportunity to create synergies between urbanization and climate change in times of crises such as the COVID-19 pandemic. The degree to which cities successfully innovate in these multi-criteria recovery efforts has a high potential to interact with and strengthen adaptation and resilience to climate change.

The COVID-19 pandemic potentially created situations of compounded vulnerability during extreme hydrometeorological events, generating a variety of direct impacts on people and the built environment (Simonovic et al., 2021; Walton et al., 2021). During a six-month period in 2020, eighty-four disasters occurred worldwide during the pandemic, including storms, floods, and droughts, impacting 51.6 million people (Romanello et al., 2021). Some examples are Cyclone Harold in April 2020, which hit the Solomon Islands, Vanuatu, Fiji, and Tonga; Typhoon Vongfong in May 2020, which struck the Philippines; and Cyclones Amphan and Nisarga that hit India and Bangladesh, respectively, in May and June 2020. Heavy floods were experienced in the Yucatan Peninsula in Mexico and Zhengzhou, China in June and July 2021, respectively (Table 1).

Compound vulnerabilities from the COVID-19 pandemic and extreme climate events and situations like relief operations for cyclones/typhoons

Table 1 Climate-related disasters in 2020–2021

Disaster	Location	Date
Cyclone Harold[a]	Solomon Islands, Vanuatu, Fiji, and Tonga	April 2020
Typhoon Vongfong[a] Typhoons Ulysses and Rolly[b]	Philippines	2020
Cyclone Amphan[a]	India	May 2020
Cyclone Nisarga[a]	Bangladesh	June 2020
Hurricane Hanna[a]	Texas, USA	July 2020
Hurricane Eta	Honduras	November 2020
Floods[a]	Yucatan Peninsula, Mexico	June 2021
Floods[a]	Zhengzhou, China	July 2021

Sources: a – Shultz et al., 2020; b – CRED & UNDRR, 2021.

amid the pandemic are not well documented. One exception is Cyclone Amphan, which triggered mass evacuation and emergency sheltering, eighty fatalities, and a spike in COVID-19 cases (Shultz et al., 2020). Extreme precipitation, floods, or cyclones led to the establishment of shelters which may or may not have included COVID-19 protocols. In July 2020, Texas, USA, experienced flash flooding from Hurricane Hanna. The state responded by organizing two categories of shelters in the areas with the highest rates of COVID-19 hospitalizations. One shelter type was designated for families that had been exposed to COVID-19 and another for the remaining people affected (Shultz et al., 2020).

Protocols during the COVID-19 pandemic also impeded the implementation of heat action plans. For example, some local governments in the USA advised residents to go to public cooling centers during extreme heat, where proximity increased the propagation of COVID-19 (Daanen et al., 2021). These recommended actions conflicted with COVID-19 directives emphasizing stay-at-home and social-distancing practices. During summer 2020, buildings identified as public cooling centers in several US cities had to shut down due to lack of mandatory air filtration within their cooling systems, which reduces the spread of airborne pathogens. These actions increased the risk of morbidity and mortality due to both heat waves and COVID-19.

Urban poor residents, particularly slum dwellers, face greater exposure to environmental hazards and climate change risks. This can increase the health risks posed by diseases like COVID-19, but also constrain the achievement of other societal objectives like SDGs (Dodman et al., 2022). For example, India experienced severe COVID-19 waves during the peak summer months of both 2020 and 2021. In cities such as Mumbai and Ahmedabad, ambient temperatures in some locations were several degrees above 40°C, due in part to the urban heat island effect (Bhide, 2021; Mahadevia et al., 2020). Families of four to five people were confined indoors in single eighty square foot tin-roofed houses (Bhide, 2021). While avoiding COVID-19 infection, residents experienced heat-related morbidity, especially those with constrained or no resources to cope with heat. More evidence on this interaction is required.

Chronic diseases related to air and water quality as well as to diet are shown to have strong positive correlations with both climate change and the COVID-19 outbreak (Bourdrel et al., 2021). Sustainable urban and peri-urban agriculture bolstered food security during COVID-19 lockdowns, for example in Dhaka (Taylor, 2020), Tokyo (Lida et al., 2023), and Mexico City (see Case Study 2).

CASE STUDY 2 MEXICO CITY'S ALTÉPETL PROGRAM: TACKLING PANDEMIC EFFECTS AND CLIMATE[9]

Gian C. Delgado Ramos

About 41 percent of Mexico City's land is urban, with the remaining 59 percent either rural or conservation land offering crucial ecosystem services including biodiversity conservation, soil stabilization, water infiltration, and CO_2 sequestration. In 2015, carbon stored in Mexico City's conservation land was estimated to exceed 3 million tons of equivalent CO_2, despite historical degradation of peri-urban ecosystems due to urbanization, illegal logging, and forest fires (SEDEMA, 2016). As a measure to reverse these trends, in 2019 the Government of Mexico City implemented a six-year program, named the Altépetl program. With 1,000 million pesos of funding per year (about 500 million US dollars), the program is focused on five areas:

(1) Protection, conservation, preservation and restoration of forest and natural resources
(2) Promoting agroforestry and silvopastoral systems by granting economic and technical assistance to production units
(3) Enhancing and expanding sustainable agricultural and livestock activities, from production to marketing, using the same mechanisms as in (2)
(4) Technical support through local capacity building ("facilitators for change")
(5) Capacity development for rural well-being through the professionalization of operational and administrative systems within the structure of the overall program.

Considering that the urban poor have been most affected by the COVID-19 pandemic and its implications, the program plays a key role in supporting the livelihoods of hundreds of families living in urban and peri-urban areas (as of 2022, more than 50,000 individuals were reported as program beneficiaries). The program supported local producers while improving their income and sustainable production capacities (grants are conditional on prohibition of the use of agrochemicals). For others, it meant a way to access local, fresh, and inexpensive food through local distribution chains, food fairs, and e-commerce (the main products, some certified as organic and GMO

---

[9] See extended version of case study at https://uccrn.ei.columbia.edu/case-studies.

free, include corn, oats, amaranth, nopal cactus, broccoli, radish, purslane, a diversity of fruits, edible mushrooms, eggs, and honey). The program was also an important source of employment during a time when unemployment grew, mainly due to COVID-19 lockdown.

From 2019 to 2022, the Altépetl program reported 19.2 million hectares of community conservation areas; 23.1 million plants and trees planted; more than 50,000 economic grants to more than 167,000 indirect beneficiaries; recovery of almost 5,000 hectares of idle land; capacity building for 15,000 facilitators for change; and the certification and promotion of dozens of locally produced products (SEDEMA, 2022).

The program has been a central policy instrument for improving well-being in rural areas where territorial identity and belonging has been eroded due to the persistence of poverty (exacerbated by the COVID-19 lockdown and subsequent socioeconomic impacts, particularly food-price inflation) and land conservation. Improving livelihoods in this population supports the preservation of ecosystem services and conservation efforts which are crucial for Mexico City, especially water infiltration, soil retention, carbon capture, and biodiversity protection.

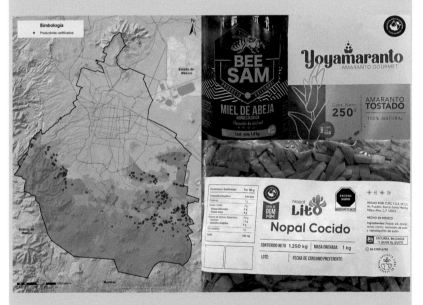

Atepetl Urban Agriculture Program for Mexico City
(Source: Delgado Ramos, 2024).

3.3 Air Quality

Cities where energy consumption declined during lockdowns usually exhibited better air quality, mainly due to reduced car use and industrial production.[10] This offers insights for emissions reduction policies and the management of related public health risks (Feng et al., 2022; Valdés & Caballero, 2020). Other studies endorse this trend, confirming that changes in air quality are statistically related to various stages, restrictions, and lockdown intensities during the pandemic (Niu et al., 2022; Vega et al., 2021).

Air pollution coupled with preexisting medical conditions can increase the chances of death from COVID-19 (Wu et al., 2020). The health effects of air pollution include cardiovascular and respiratory disease, cancer, stunted brain development, negative birth outcomes, and more (Khomenko et al., 2021; WHO, 2016). These may be exacerbated by a diversity of preexisting medical conditions related to unhealthy diets and lack of exercise (Guthold et al., 2018; Norris et al., 2022). Air pollution may accelerate virus propagation while also contributing to climate change (Coccia, 2020).[11] For example, in Chinese cities, there was a significant positive association between long-term exposure to the particulate matters (PMs) $PM_{2.5}$, PM_{10}, NO_2, and Oxone (O_3) and the risk and severity of COVID-19 infections (Wang et al., 2020; Yao et al., 2021). In twenty densely populated Indian cities, COVID-19 seemed to spread faster in high temperatures, especially at temperature and humidity that ranged between 27 and 32°C and between 25 and 45 percent, respectively (Sasikumar et al., 2020).

In Europe, a study of sixty-six administrative regions in Italy, Spain, France, and Germany found that 78 percent of COVID-19 deaths at March 19, 2020 took place in five urbanized regions located in north Italy and central Spain, all of which had high nitrogen dioxide (NO_2) concentrations and downward airflows that prevent air pollution dispersion (Ogen, 2020). A similar outcome was reported by seventy-one Italian provinces (Fattorini & Regoli, 2020). In the Netherlands, COVID-19 cases, hospitalizations, and deaths were positively correlated with high $PM_{2.5}$ concentrations (Cole et al., 2020). In England, correlations were calculated for long-term exposure to fine particulates and showed an increase of up to 7 percent in COVID-19 spread and mortality (Office for National Statistics, 2020).

---

[10] Not all cities experienced improved air quality during the COVID-19 pandemic. Santiago and Mexico City did not show significant reductions in air pollutant concentrations (León & Cárdenas, 2020). Despite the slowdown of economic activities in Mexico City, nearby forest fires likely caused poor air quality, contributing 79–92 percent of primary fine particle mass to the metropolitan area (Yokelson et al., 2007).

[11] See ARC3.3 Element on *Urban Planning, Design, and Architecture*, www.cambridge.org/core/publications/elements/elements-in-climate-change-and-cities.

Similarly, in a study of twenty-five Mexican cities, a 3.5 percent increment in COVID-19 mortality rates was found for every additional 1μg/m^3 of NO$_2$ (Cabrera-Cano et al., 2021). This has been confirmed by a significant positive association between PM$_{2.5}$, carbon monoxide (CO), and O$_3$ and COVID-19 infections and deaths that was found in another study carried out in Mexico City between April 1 and May 31, 2020 (Kutralam-Muniasamy et al., 2021). The association between increases in COVID-19 disease and long-term exposure to air pollutants has also been reported in twenty-four districts in Lima, Peru (Vasquez-Apestegui et al., 2021), while in Brazil, a 1μg/m^3 of PM$_{2.5}$ was associated with a 10.22 percent increase in COVID-19 deaths (Damasceno et al., 2023).

Research generally suggests that the outdoor environment affects the propagation of COVID-19 as well as other respiratory diseases through factors like air pollution, limited urban ventilation, high temperature and humidity, and lack of green infrastructure. Indoor air pollution, primarily due to the use of residential gas stoves, has also been associated with adverse health effects, potentially aggravating COVID-19 impacts (Piscitelli et al., 2022; WHO, 2021b). Therefore, improving outdoor and indoor air quality has multiple benefits, whether related to climate, the environment, or public and individual health.

4 Urban Systems: Built Environment, Transportation, and Waste

Cities have begun to undertake built-environment development for climate change adaptation and mitigation. For example, to reduce pollution from cars and improve community well-being, New York City turned Times Square from a circulation nightmare into a pedestrian-safe gathering space for events to take place and for people to enjoy. Paris started removing car lanes and replacing them with bike and pedestrian lanes. Parking spots became seating areas, and either sections or entire streets have been closed off to cars and opened up for bicycles and pedestrians (Yassin, 2019).

Proposals for reevaluating streets have emerged – for example, from the perspective of "complete streets" which seeks to promote active streetscapes, green infrastructure, street equipment (including pedestrian-scale lighting, signage, and bicycle facilities), public transportation, and accessibility and mobility for all through introduction of median islands, curb extensions, and wider sidewalks. Complete street initiatives have been implemented in cities in the USA, Canada, Mexico, Brazil, and India, among other countries.

These initiatives were enhanced during the COVID-19 pandemic, particularly during the first waves. For example, banning cars on streets was observed under Open Streets programs – for example, in New York City (Finn, 2020) and Barcelona[12] – so that street spaces could better serve businesses and their customers and provide more areas for walking and cycling (Combs & Pardo, 2021). These programs allowed restaurants to temporarily extend their seating space into public areas, increasing capacity while also enabling social distancing and limiting indoor exposure to COVID-19. The long-term operation of Open Streets programs is unclear; in some cities, such changes in usage of the built environment are still continuing, while in others business-as-usual street configurations have returned. (See the additional resources for more information about establishing cycling lanes in Monterrey City, Mexico.)

4.1 Sprawl and Market Demand

In response to climate change agendas, cities have implemented inner-city redevelopment and regeneration programs with the aim of reducing poverty in these areas and curtailing sprawl, thus advancing resilience as well as reducing GHG emissions. Cities have also implemented transit-oriented development (TOD) – high-density developments along transport corridors, with the aim of concentrating urban populations (Jaramillo et al., 2022). However, a misconception that high-density areas were more prone to the spread of COVID-19, the fear of infection in elevators, and the implementation of long and unclear lockdown measures pushed households who could afford it into low-density peripheral locations and larger homes with more access to green and open spaces (Mouratidis, 2022).

Urban sprawl is not a recent phenomenon, but a process that has occurred over the last thirty years (Güneralp et al., 2020; Liu et al., 2020; Mahtta et al., 2019). It not only increases demand for travel and carbon emissions, but it also destabilizes surrounding natural areas through expansion of impervious surfaces that exacerbate urban flooding and heat-island effects.

In cities in the USA, people from large cities started to choose locations in more affordable suburbs, smaller cities, and areas away from high-cost, high-density urban downtowns (Li & Zhang, 2021). This trend has reversed the "back to the city movement" that encourages people to shift from the city's outskirts to older, more central neighborhoods, often through inner-city regeneration programs over the last two to three decades. Some cities, such as Boise, Idaho and Austin, Texas experienced in-migration, as documented through property purchases and rentals during the COVID-19 pandemic. This was accompanied by

---

[12] See New York City DOT, www.nyc.gov/html/dot/html/pedestrians/openstreets.shtml (accessed May 19, 2023) and The Mayor.EU, www.themayor.eu/en/a/view/open-streets-returns-to-barcelona-in-march-4309 (accessed May 19, 2023).

a loss of population in highly urbanized counties such as Los Angeles, Chicago, and New York, where resident outflows have been estimated to be between 102,000 and 185,000 people from April 2020 to July 2021 (Prater & Lichtenberg, 2022), and up to 300,000 in these US cities (US Census Bureau, 2022). However, the US National Apartment Association and others have pointed to a positive outcome of this process, calling it a "shuffle" rather than an exodus. As residents moved to more livable cities or rural areas during the pandemic, rental prices eventually started to decline, making way for inbound residents to fill those vacant properties (NAA, 2021; Whitaker, 2021). Move-ins in newer locations are being attracted by lower rents, or similar rents for larger homes as compared to inner city properties, as well as by the now more-accepted possibility of working remotely.

A similar process was reported in London neighborhoods, where an increase in prices for detached, semi-detached, and terraced houses and a decrease in prices for flats and maisonettes, a phenomenon associated with a growing demand for living and work-from-home spaces, has been described (Cheshire et al., 2021). The decentralization of urbanization trends to rural, peri-urban, and small towns may not entirely conflict with climate change mitigation efforts. If planned correctly, this could become an opportunity to repopulate and regenerate rural areas. Moving to a sustainable lifestyle through lower use of transport, especially in the context of possibilities of working from home, in combination with the use of renewable energy could reduce household GHG emissions. However, repopulating rural areas or intensifying urban population density may not support a desirable transformation pathway as this simultaneously depends on other factors, from land use to urban form and structure. (See the additional resources for more accounts of COVID-19 related urban sprawl.)

Accelerating urban sprawl, with associated mitigation setbacks, is a global phenomenon. After lockdowns and the relaxation of travel measures, a "digital nomadism" started to be more evident in some places, such as Mexico City, where foreigners started to migrate, profiting from cheaper rentals while working remotely with a salary in dollars or euros. This digital nomadism, mostly an outcome of the COVID-19 pandemic, eventually pushed up rental prices in central city areas – like the Roma and Condesa boroughs in Mexico City – exacerbating a gentrification process that has forced residents to relocate to other boroughs and even to nearby cities with cheaper rentals (Ware & Mariwany, 2022). A similar process of digital nomadism has also been reported in Spain (Parreño-Castellano et al., 2022) and other locations, mostly in the Global South (Holleran, 2022), a context in which most of these relocations are short-lived or have lingering consequences for local housing markets. This phenomenon shows how urban dynamics are frequently interconnected, globally. Therefore, effective solutions need local-to-global coordinated actions, even more so when shuffle or short-distance migration/exodus dynamics seem to constitute two sides of the same contradictory process, as experienced in the USA (PWC & Urban Land Institute, 2022).

4.2 Transportation Management, Travel Patterns, and Mobility

Safety is crucial for public transit systems. The notion of safety was restricted to safety against violence and crime, such as homicide, mugging, sexual harassment of women, and racial incidents, but has been expanded to include safety from infection. Connectivity and mobility may be at the forefront of the causes of the spread of COVID-19 (WHO, 2020b) because transmission has been found to increase indoors (within shared transport) compared to outdoors (Rowe et al., 2021). Increasing risk of virus contagion in shared travel modes raised fears (De Vos, 2020; Fundación Gonzalo Rodríguez & OPS, 2022) which impacted patterns of travel activity either as a consequence of COVID-19 containment measures or through personal choice. As a result, public transit use decreased during COVID-19 transmission peaks (see the additional resources, Table 1).

This widespread trend had financial implications due to declining revenues and increasing expenditures (such as those related to hygiene procedures, the purchase of personal protective equipment (PPE) kits, and increased services to comply with physical distancing). It became more difficult to invest in sustainable transport systems when public funds for transport were being reduced (Olin, 2020). For some, public transport is the only option for commuting: if they were unable or unwilling to walk, cycle, or drive a car, their livelihoods were severely impacted (Vitrano, 2021).

> CASE STUDY 3 EXACERBATING THE FINANCIAL WOES OF TRANSPORT FOR LONDON: A LONG-TERM IMPACT OF COVID-19[13]
>
> *Darshini Mahadevia, Saumya Lathia, and Amitkumar Dubey*
>
> Since Transport for London (TfL) was founded in 2000, daily public transit trips in the UK's capital have matched the rest of the country. The public body has successfully increased public transport usage by 65 percent and decreased carbon emissions by 7 percent between 2000 and 2018, despite increased demand. It had supported 43,000 jobs around the UK and contributed £7 billion to the UK economy by 2018. However, since 2010, TfL's operating grants from central government have been reduced and, in 2018, experts predicted its bankruptcy should government assistance continue to

---
[13] See extended version of case study at https://uccrn.ei.columbia.edu/case-studies.

decline. The COVID-19 pandemic exacerbated TfL's financial stress and led to a catastrophic decline in the use of public transport.

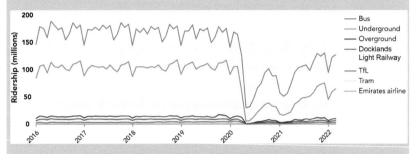

Change in TfL ridership due to COVID-19
(Transport for London dataset, https://data.london.gov.uk/dataset/public-transport-journeys-type-transport, accessed April 29, 2022).

On March 18, 2020, the Mayor of London declared a discontinuation of all "nonessential" activities, fueling a 92 percent decrease in transit use and a subsequent loss of £500 million, 72 percent of its total revenue. Although TfL had one of the most comprehensive coping mechanisms during the COVID-19 pandemic, prioritizing the safety and accessibility of its users, its unsustainable revenue model (passenger revenue contributed about 72 percent of TfL's total income) worsened during the pandemic. Even with a reduction in operating costs, TfL could not compensate for the financial losses of lost passengers.

On May 15, 2020, after several negotiations with central government, the first round of financial support allowed TfL to function at 70 percent capacity. In accordance with TfL's projections pre-pandemic, it would have continued to incur debt until 2022, after which it would began generating a surplus (see the additional resources for more information). Due to the COVID-19 pandemic, the steep decline in passengers derailed its financial plans of preventing bankruptcy.

This case study highlights the need for governments to evaluate public transport systems as a necessary utility that enables access to socioeconomic opportunities and growth. Climate change is likely to increase future risks to health as well as disruption to public transport services due to extreme weather events. Therefore, at the local scale, cities should (re)build resilient public transport, and national governments need to step in to meet revenue shortfalls as a result of disruptions like COVID-19.

To reduce the spread of COVID-19 through public transport, diverse measures were adopted by local governments (Patlins, 2021). These included:

- Operating public transport on select routes only (Czödörová et al., 2021)
- Running public transport buses only during peak times with crowding management strategies (Gkiotsalitis & Cats, 2021)
- Implementing antiepidemic measures like reduced occupancy and social distancing (Hörcher et al., 2022; Pardo et al., 2021)
- Frequently disinfecting vehicles, especially surfaces that people touch like handgrips, door controls, and handles (IAPT, 2020)
- Providing hand sanitizers for passengers (American Public Transportation Association, 2020; UN-Habitat, 2020b).

In cities with poor public transport and intermediate public transport (such as auto-rickshaws in Indian cities), certain safety measures were put in place. For example, in Ahmedabad, India, a rule to limit occupancy of auto-rickshaws to two passengers (instead of 3–6) was enforced, which increased the price of journeys (Indian Express, 2020). Similar measures have been reported for car-sharing services, which verified a decrease in demand during pandemic peaks (Garaus & Garaus, 2021). By enforcing the use of face masks and a multistep safety-screen process for drivers, among other actions, car-sharing platforms have the potential to reduce not just car ownership but kilometers traveled and ultimately GHG emissions – as reported in the Netherlands (Nijland & van Meerkerk, 2017).

The additional expense of putting in place safety measures during the COVID-19 outbreak when public transit use declined created financial losses for transit companies (see Case Study 3). Some public transit companies in US cities were already financially strained, a situation that was exacerbated by the pandemic and has led to stagnant wages for transit workers, resulting in an overall shortage of operators and mechanics in these companies (American Public Transportation Association, 2023).

Despite these measures, residents of many cities preferred, where possible, to use private motorized transport, cycling, or walking (Das et al., 2021; Zhang et al., 2021). As a result, automobile ownership has increased in some cities, thus requiring more parking spaces (Basu & Ferreira, 2021) which many cities had been trying to reduce as part of mitigation efforts to lower GHG emissions. The shift to private vehicle use was reported in European and Canadian cities (see the additional resources for more information).

These cases demonstrate that perceptions of risk from COVID-19 and any future pandemic have potential to alter travel behavior, with people making choices to shift to private mobility. Planning of public transport systems to increase resilience to contagious diseases is required if low-carbon transport options are to be supported. While initial trends show a decrease in overall global emissions due to lockdowns during the pandemic within the range 5.4–5.8 percent (UNEP, 2021), return to business as usual in concert with greater automobile ownership may have longer-term negative consequences (Das et al., 2021). Rebounding transport activity after COVID-19 restrictions had been lifted led to an 8 percent increase in CO_2 emissions from transport between 2020 and 2021 (IEA, 2022a).

Multiple sustainable transportation and mobility options are available going forward and learning from the COVID-19 pandemic. Three key options are described here: (1) promoting mixed land use as the core of urban intensification efforts that can lead to a reduction in travel distances; (2) repurposing urban public spaces so as to advance active mobility; and (3) supporting remote working.

The first option is change in land use that creates access to everyday services within walking distance by developing mixed land use together with the consolidation of well-connected and vibrant urban centers and subcenters. For example, Paris promotes the "15-minute city" (Moreno et al., 2021), enabling an agenda for coping with global climate change while strengthening public health and well-being (see Case Study 4). More research is required on the success of this approach. Travel restrictions in the Netherlands led to the formation of "activity bubbles" in which residents permanently become more reliant on the amenities available in their home district (Champlin et al., 2023). However, bridging social interactions across district boundaries, or the formation of larger activity bubbles, is associated with higher social resilience and well-being.

CASE STUDY 4 IMPLEMENTING THE "15-MINUTE CITY": A CASE STUDY OF PARIS[14]

Carlos Moreno, Zaheer Allam, and Didier Chabaud

The 15-minute city model has been proposed as a path to urban sustainability and resilience; it is recognized as a way forward for economic regeneration that can support quality of life. Implemented by the city of Paris as a policy priority axis by Mayor Anne Hidalgo (2020–2026), the model has become a new urban planning framework based on favoring proximity, diversity, density, and digitalization while offering climate justice and socioeconomic equity. To carry out this strategy articulated around new hyperproximity services, an in-depth transformation of the Parisian administration has been required to empower neighborhoods for regenerative action, backed with appropriate resources.

---

[14] See extended version of case study at https://uccrn.ei.columbia.edu/case-studies.

Examples of projects being developed include playgrounds open to residents, open-air libraries, schoolyards and colleges transformed into "oases," pedestrianization of districts and the creation of school streets to provide security around schools and promote reduced carbon emissions, and interventions to embellish neighborhoods in conjunction with residents. Citizen participation has been promoted – for example, through participatory budgeting and the establishment of a volunteer program and citizens' kiosks offering information on municipal action and helping to articulate citizens' proposals and actions. Paris has also launched a new bioclimatic local urban plan that is intended to provide some autonomy in deciding its housing and urban revitalization development plans.

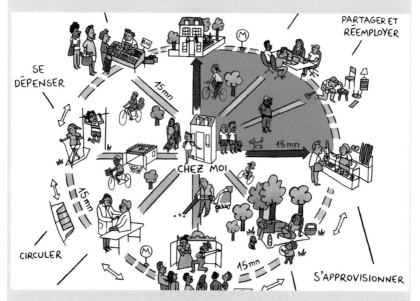

15-minute city (Source: Micael in Paris en Commun).

Similar 15- to 20-minute city strategies are being considered and implemented elsewhere, for example in Milan (Pisano, 2020), Liverpool, UK (Calafiore et al., 2022), Portland, Oregon (Simon, 2022), and Melbourne (Victoria State Government, 2019). The model is gaining more attention in the context of planning post-COVID-19 cities.

The second option is repurposing – and expanding – urban public spaces to create safe paths for active mobility. More than 500 cities, states, and countries did so during the COVID-19 pandemic, between March and August 2020 (Combs & Pardo, 2021) (Figure 6). In some cases, such street-space adaptations, in addition to measures to adjust mobility demand and behavior, have since been

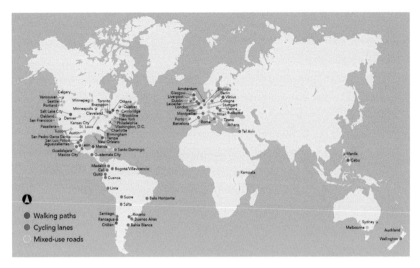

Figure 6 Sample of cities that implemented infrastructure such as walking paths, cycling lanes, and mixed-use roads for active mobility in 2020 (IDB, 2020).

coupled with green/blue infrastructure. Other measures being taken are expansion of cycling lanes and bicycle parking; closure of streets to motor vehicles; reduced speed limits; automated walk signs; subsidized bicycle purchases, repair and sharing systems; shared streets; and reallocation of curb space, roadway, or nonstreet spaces to walking, cycling, or outdoor commerce (Combs & Pardo, 2021). Examples of land-use repurposing are largely from Europe, North America, and Latin America (see the additional resources, Table 2).

During the COVID-19 pandemic cycling rates increased in many cities in comparison to other modes of transport. For example, average cycling rates increased by 27 percent in Italy, 23 percent in the UK, 19 percent in Portugal, 16 percent in Spain, and 15 percent in France between 2019 and 2021 (Buehler & Pucher, 2022). Expectations that cycling rates would increase were reported for Seoul (Ku et al., 2021) and Dutch cities, which already had high cycling rates. In the latter case, and because of newly formed habits during the COVID-19 emergency, 20 percent of people are expected to cycle and walk more in the future (de Haas et al., 2020). (For further details, see Castro-Sánchez [2024]; also see the additional resources for more on cycling during the pandemic.)

The third option is to support remote working through the continued use of massive, newly developed teleworking infrastructure (Edelman & Millard, 2021). Some companies continue to allow employees to work part-time from home, which benefits companies financially through reduced need for office space. While under certain conditions this option has (Guerin, 2021) positive implications for reducing GHG emissions from transport and improving local air quality, it has the potential not only to create challenges for work–life balance (Atkinson, 2022), but also to increase residential energy use (Villeneuve et al., 2021). Energy consumption from data centers reached 220–320 terawatt hours (TWh) or up to 1.3 percent of global electricity demand in 2021 (IEA, 2022b).

4.3 Circularity and Urban Waste Management

Circularity in waste management, including building components and construction and demolition waste, can contribute to sustainability and climate change mitigation beyond conventional recycling and composting practices (Hertwich et al., 2020) and has recently begun to be implemented in practice. The COVID-19 pandemic disrupted local municipal services including formal and informal waste management at the local level, particularly during the first wave (Bel & Marengo, 2020; Roy et al., 2021). Larger amounts of single-used plastics, face masks and other PPE, as well as packaging waste – due to changes in consumer buying patterns – increased the already limited and disrupted capacities of some municipal waste services, while also undermining efforts to reduce plastic waste (Leal Filho et al., 2021).

The surge in reusable face mask production (Corrêa & Corrêa, 2021) led to higher use of plastics. Pandemic-associated plastic waste generated by 193 countries, as of August 23, 2021, has been estimated at 8.4 ± 1.4 million tons of which 25.9 ± 3.8 thousand tons have already entered global oceans; this represents 1.5 ± 0.2 percent of total riverine plastic discharge globally (Peng et al., 2021). The use of plastics in face masks and other protective equipment production, in 2020, represented a volume of ~ 300,000 tons (OECD, 2022a). Due to disease spread, lockdowns, and constrained financial capacities, the circular economies of cities were adversely affected (Rahman et al., 2021; Roberts & Drake, 2021).

Circular economy strategies that can be applied to prevent and reduce waste generation, as well as to manage waste may help reboot the labor market after the COVID-19 pandemic by engaging unemployed people in circular jobs. Working with informal communities on waste management can promote circularity while improving the livelihoods of waste pickers and other informal waste workers (Ellen MacArthur Foundation, 2020; Sharma et al., 2021).

The COVID-19 crisis illustrates, on the one hand, opportunities and synergies between climate, environmental, and public health agendas and, on the other, multiscalar disruptions in both metabolic inflows (challenges in the supply chains of strategic products) and outflows (challenges in managing waste and wastewater during the pandemic). The crisis foregrounded short-term opportunities, such as urban and close-to-city agriculture, local sustainable reusable-mask production, and goods and restaurant food delivery, among others. However, the impacts on longer supply chains were often much greater.

The concept of circular cities becomes useful in transforming many facets of urban dynamics while presenting opportunities to reconnect cities to nearby rural countrysides and develop new economic models (UNEP & UN-Habitat, 2021). Nevertheless, circular economy models need to resolve not just technological and managerial issues, but also economic, regulatory, and culturally localized challenges or bottlenecks (Montag et al., 2021; Newman, 2020).

5 Interactions with Urban Ecology

Cities around the world have begun to introduce nature-based solutions into infrastructure and management systems (Dodman et al., 2022; Lwasa & Seto, 2022).[15] Access to green spaces was associated with a lower spread of COVID-19 in several cities throughout the world as well as with a reduced risk of COVID-19 mortality (Hong and Choi, 2021; Russette et al., 2021).

Urban nature-based solutions, such as access to green spaces, offer benefits to city ecosystems as well as to the physical and mental health of humans (Kabisch et al., 2017). In a study of 135 highly urbanized counties in the USA, the presence of green spaces was significantly correlated with lower racial disparity in the COVID-19 infection rate (Chen et al., 2020). Another study of forty-five countries showed that in places where parks were kept open, visitors increased during the COVID-19 pandemic compared to the numbers visiting before (Geng et al., 2021). In most of the countries analyzed, there was a significant negative correlation between the number of park visitors and the daily increase in COVID-19 cases, except in Finland and Sweden where there was a significant positive correlation. During COVID-19 lockdowns and government stringency, there was a negative association with park visiting, as expected.

In another study, survey respondents in Buenos Aires described urban green space as "a place to be with nature" before confinement and "an important place in the city" afterward, reflecting the growing importance of green spaces for urban residents post-pandemic (Marconi et al., 2022). In Oslo, Norway, outdoor recreational activity during lockdown increased by 291 percent, with the greatest increase in greener and more remote parts of the city. This increase was sustained for months after the COVID-19 outbreak (Venter et al., 2020).

In the Asian cities of Hong Kong, Singapore, Tokyo, and Seoul, researchers recorded a 5.3 percent increase in the probability of people using green spaces for every weekly 100-case increase, indicating that more urban dwellers were frequenting parks as COVID-19 cases rose (Figure 7). This study used social

---

[15] See ARC3.3 Element on *Nature-Based Solutions* for an in-depth assessment of this topic, www.cambridge.org/core/publications/elements/elements-in-climate-change-and-cities.

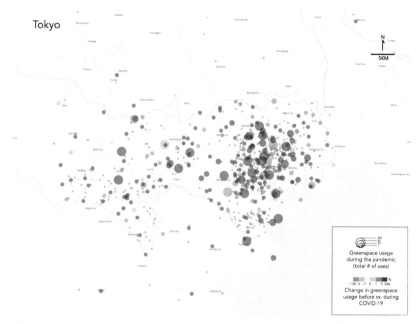

Figure 7 Change in green space usage during the pandemic in Tokyo. Circle size represents the total number of green space uses; color represents change of use during versus before the pandemic (Lu et al., 2021). Park use sample period is from December 16, 2019–March 29, 2020, and total # of uses of individual parks is tracked from Instagram data.

media data to show that people preferred large, natural parks that were close to city centers (e.g., nature trails) as opposed to urban parks with man-made elements (e.g., football fields) (Lu et al., 2021).

Decreased green space usage and park visiting were also recorded during the COVID-19 pandemic. Closure of green spaces during the early stages of the pandemic due to the perceived risk of COVID-19 transmission may have resulted in a net-negative effect on community health. A study conducted in South Korea found that 64.9 percent of survey respondents reported decreased visits to green spaces after the COVID-19 outbreak (Heo et al., 2021).

In Croatia, Israel, Italy, Lithuania, Slovenia, and Spain, urban residents visited green spaces less during the containment period, with a relative increase in necessary activities such as taking the dog out and a relative decrease in nonessential activities such as meeting others (Ugolini et al., 2020). Residents of UK cities also reported fewer visits to green spaces due to movement restrictions (Burnett et al., 2021). In Brisbane, Australia, observed behavior was mixed: 36 percent of respondents increased their visits to parks while 26 percent reduced them (Berdejo-Espinola et al., 2021).

The availability of, and hence inequitable access to, green spaces in some cities was and still is an issue (see the additional resources for city examples). Overall, urban green spaces played an important role in mental and physical health prior to the pandemic as well as during lockdowns and subsequent COVID-19 waves. Nature-based solutions also have a critical role in climate change mitigation, adaptation, and resilience and offer immense benefits to urban populations and ecosystems. (See the additional resources for impacts on urban and peri-urban wildlife.)

6 Governance and Urban Climate Action

The intertwined nature of climate change adaptation and mitigation and pandemic preparedness in cities requires coordinating actors, institutions, jurisdictions, knowledge providers, agencies, and practices at different scales, both in time and space. This type of governance,[16] needed to move toward more sustainable, healthy, resilient, and just cities, may be understood as a reflexive way to build consensus and partnerships based on scientific information and knowledge coproduction around the definition of goals and pathways of action in fragmented and uncertain environments (Solecki et al., 2021; UN-Habitat, 2022).

Governance challenges emerge when there is a lack of alignment or even mismatch across scales, interdependence between levels, or "contagion" or transfer of responsibilities from one level to another (Termeer et al., 2019). Transfer of responsibilities could be explained as, for example, taking over pandemic management by the national government or, vice versa, passing down financial responsibility for health care during the pandemic. Solutions require the development of better links between levels to address mismatches between the scale of the problems and the scale at which they are governed (Cash et al., 2006). In this context, leadership is not only crucial to ensure rapid and appropriate responses, but to catalyze transformative actions and unlock deadlocks in governance processes (Termeer et al., 2019).

Solutions also require an understanding of a wide range of social phenomena, such as the evolution of political institutions and of informal governance practices, the progression of economic development and financial markets, as well as disaster prevention, response, and preparedness (Bai et al., 2020). Solutions need to be local, regional, and global, as evidenced in the COVID-19 pandemic, where international collaboration in financial support, donation of medical supplies, and the sharing of expertise occurred (Dalglish, 2020; Mahmud & Al-Mohaimeed, 2020.

---

[16] See ARC3.3 Element on *Governance, Enabling Policy Environments, and Just Transitions*, www.cambridge.org/core/publications/elements/elements-in-climate-change-and-cities.

6.1 Multilevel Governance and City Responses

Governance during the COVID-19 pandemic offers lessons for institutionalizing resilience to climate change impacts as well as for putting in place climate change adaptation and mitigation efforts in the short and the long run. Building economic, social, and climate–environmental resilience and the appropriate governance and institutional structures "must be at the heart of the future of cities" (UN-Habitat, 2022). To achieve this, however, a consistent and coordinated agenda and the means to implement it needs to be developed at all levels. The COVID-19 crisis has certainly been an opportunity to reveal challenges and identify opportunities for advancing strong multilevel governance for low-carbon, sustainable, and resilient urban transformation.

Governments – national, provincial, and local – developed policies to contain the spread of COVID-19 and provide healthcare to those afflicted. The responses, from neighborhood to subnational, national, and supranational scales, varied depending on different state as well as civil society capacities. There was a range of efficacy in providing relief during lockdown measures to contain the spread of the disease, provide healthcare, and promulgate interventions for recovery after localized waves of infections.

Initial analyses suggested that the COVID-19 pandemic led to increased authoritarianism (Gao & Zhang, 2021; Lührmann et al., 2020). The pandemic resulted in widespread isolation, strict lockdowns, and policy-enforced quarantines in many places, despite a decades-old acknowledgment that oppressive public health measures are considered counterproductive (Bohle et al., 2022; Hartman et al., 2021; Holbig, 2022). The wrong lessons could be drawn for climate change actions – such urgent mitigation measures to meet the challenge of a 1.5 °C temperature increase – from COVID-19 pandemic management through authoritarian practices.

Early responses to the pandemic in many countries was for the national government to take over decision-making that would otherwise have been the domain of provincial or local governments, such as decisions related to health. In some countries, economic development and utilities sectors, reinforced centralization practices. The COVID-19 emergency revealed a tendency to centralize (UCLG et al., 2021). Decentralization processes were observed more often in containment and food security measures, while transport included shifts in both directions – centralization and decentralization practices (Dzigbede et al., 2020; UCLG et al., 2021). In general, the COVID-19 emergency elicited a range of actions, mostly in relation to emergency preparedness and response, information and communication technology, and the concentration of executive powers

(UCLG et al., 2021). Figure 8 provides an example of US Center for Disease Control and Prevention (CDC) communication measures on hygiene and handwashing in Times Square New York. Local governments that were resource-poor were not able to respond well to the pandemic (Dzigbede et al., 2020), while those with better resources responded with mixed results.

Figure 8 Empty Times Square in New York with billboard slogan: "Wash hands often with soap & water for at least 20 sec."
(Source: Erik Mencos Contreras, March 2020).

As in the climate change adaptation literature, studies published on COVID-19 governance show the importance of collaboration among levels of government and nongovernmental stakeholders.[17] For instance, a comparison of the governance mechanisms in place during the pandemic in China and in the USA demonstrated that, despite different political systems, the two countries set up hybrid coordination regimes in which interactions can be found between different levels of government (vertical approach) and with stakeholders and civil society (horizontal approach) (Liu et al., 2021). In both cases, several different modes of coordination were used: command and control, steering, negotiating, and supporting (although differences are observed between the two countries). (See the additional resources for city examples.)

As a central component of multilevel urban governance, coordination includes the metropolitan scale as well as cooperation among transboundary cities, such as metropolitan agglomerations with multiple local governments but operating as unified economic entities. The COVID-19 pandemic has reaffirmed the importance of metropolitan-level coordination and transboundary cooperation within multiple local governments which, when it takes place, typically has been limited to informal or soft coordination practices that do not necessarily lead to the most effective responses (UCLG et al., 2021).

Nonstatutory agreements among local governments within metropolitan agglomerations may be on issues of taxation, land-use regulations, pollution standards, ecological resource use and deployment, building codes. Large metropolitan areas in particular reveal great disparities and inequities, not only in health outcomes or access to healthcare but also in terms of climate vulnerability and risk exposure. Metropolitan scale coordination seems to be more effective than city to city coordination (Delgado Ramos et al., 2019; Delgado Ramos & Mac Gregor Gaona, 2020).

Cities in the Global South are at more risk from both climate change and pandemic impacts due to the expected increase in urbanization and the growth of informal settlements in vulnerable and nonplanned areas (Bele et al., 2014; Field & Barros, 2014). Moving forward, institutionalization of urban planning to cope with multiple crises will be increasingly relevant.[18] Working at different spatial and time scales while integrating a set of urgent urban agendas is required to address diversity, equity, and inclusion among individuals, institutions, neighborhoods, cities, and metropolitan areas. This will enable more

---

[17] See ARC3.3 Element on *Governance, Enabling Policy Environments, and Just Transitions*, www.cambridge.org/core/publications/elements/elements-in-climate-change-and-cities.

[18] See ARC3.3 Element on *Urban Planning, Design, and Architecture*, www.cambridge.org/core/publications/elements/elements-in-climate-change-and-cities.

equitable and inclusive planning and decision-making processes. It also recognizes varying conditions and capacities to act.

6.2 Leadership, Information, Social Partnerships, and Accountability

Beyond coordination between different levels of government (see Section 6.1), creating spaces for collaboration and cooperation with citizens and stakeholders to foster knowledge sharing is important (Garavaglia et al., 2021). Leadership at the city level is also important for this purpose. A survey of twenty-five mayors in the metropolitan area of Milan found that the emergence of several public actors – volunteers, public officials, policemen, and so on – significantly helped create shared leadership to collectively manage the health crisis without weakening the overall leadership position exercised by Italian mayors (Garavaglia et al., 2021). This shared leadership is particularly important to maintain now that the pandemic is officially over. The issue of effective response and the quality of leadership also emerged during the pandemic crisis. Effective response includes rapid action, good coordination, an evidence-based approach that is well communicated, and a spirit of partnership (Al Saidi et al., 2020).

Information and communication technologies proved to be useful during the COVID-19 pandemic and its consequences – to strengthen institutional governance (Clement et al., 2023; UNDESA, 2022), advance community engagement (Spear et al., 2020), create citizen support networks (e.g., the Paraguayan initiative, ayudapy.org), and articulate social mobilization (Duque Franco et al., 2020). Participatory approaches to data collection can afford residents greater agency in the cocreation of urban data while emphasizing privacy over precision (Champlin et al., 2023). Yet, such roles and potentialities still have limitations due to the prevailing digital divide between rich and poor, young and elderly, and rural and urban environments (BEREC, 2021; Li, 2022).

Nevertheless, e-government strategies can improve capacity to respond to crises and generate useful information in real time for decision-making, as the COVID-19 crisis proved (UNDESA, 2022). Digital platforms and social media can be used effectively used to spread information and instructions and confront "infodemia" (the misinformation epidemic) (Khan et al., 2022; Onyango and Ondiek, 2022). Information and communication technologies can also be used by grassroots movements to articulate support for those in need, as demonstrated by the *ollas populares* ("popular pots") initiative led by women from vulnerable communities in Asuncion, Uruguay, who were facing food insecurity aggravated by the pandemic (Frutos et al., 2022). A similar account relating to ICT was reported in Lima (Desmaison et al., 2022).

Information and communication technologies can also be used, on the one hand, to shield governments from blame for the pandemic (as argued in the case of China, Li et al., 2022), while, on the other, they can create public trust, an important aspect of strict enforcement. (See the additional resources for city examples in Wuhan, Singapore, and Kerala.)

Social media, mobile applications, and other digital tools can be valuable when used with good intention and mindful data storage that does not infringe the privacy and rights of citizens. Social media and digital platforms are used tp give warnings in climate change-induced extreme events such as cyclones and extreme heat, as in the case of Ahmedabad's Heat Action Plan (Mahadevia et al., 2020; Nastar, 2020). The use of ICTs was likely accepted during the pandemic due to the urgent and emergency nature of the situation. However, it is also known that ICTs can be intrusive and violate individuals' privacy (Cong, 2021). The question is whether ICTs can be useful tools for decentralized and participatory urban governance, both in climate action and more generally. This is certainly an issue that needs to be further evaluated, especially given the rapid evolution of artificial intelligence.

6.3 Role of Technology in Emergency Planning and Management

The COVID-19 pandemic has not only led cities to coordinate actions across different sectors and stakeholders and incorporate ICTs to generate and disseminate information, but also to find solutions (Kummitha, 2020; Kunzmann, 2020). Learning from such experiences, urban adaptation and mitigation practices can be reinforced by fostering pre-event and long-term urban planning, and early warning for multihazard situations, as well as informed emergency plans, while advancing collaborative practices and alliances and the incorporation of novel technologies and technological abilities (Han et al., 2021).

Due to the necessity of deploying technologies in the context of COVID-19 – for tracking, understanding, and informing the spread of the disease in real time, for encouraging social distancing, or arranging vaccination campaigns and certification – digitized or smart cities are now better prepared to respond to future disasters. Urban governments gained valuable experience during the pandemic by ensuring the well-being of urban residents while maintaining continuity of urban functions (Hassankhani et al., 2021).

Despite the need for further learning on the influence of digitalization on pandemic responses, it seems evident that countries and cities with higher digital adoption highlighted a positive trend in how they handled the COVID-19 pandemic and implemented institutional interventions (Heinrichs et al., 2022; Liu et al., 2022). Therefore, the implementation of different technology-driven policies and actions seems to be desirable for assisting telehealth – for both physical and mental health (Baudier et al., 2023), advancing e-learning, remote working, and services (WEF, 2022), preventing and managing risk (Petrova & Tairov, 2022), strengthening sustainable and resilient urban fabrics (Allam et al., 2022; Cavada, 2023), and even for improving social participation and connectedness.

In Helsinki, for example, digitalization efforts, which are not centered on technology alone, have been explicitly proposed as part of the city's recovery plan, which also comprises broader actions for green structural change in the economy and the introduction of more innovative and efficient public services that support a reduction in carbon emissions (City of Helsinki, 2020). A similar proposal can be found in Los Angeles' SmartLA 2028 plan, which proposes actions relating to infrastructure, data tools and practices, digital services and applications, connectivity and digital inclusion, and city governance (City of Los Angeles, 2020). Other cities working in a similar direction include Toronto (Toronto Government, 2023), Portland (City of Portland, 2020), and Boston (City of Boston, 2022).

7 Energy and Economics

The COVID-19 pandemic revealed underinvestment in public health, but also neglected climate commitments at the local level due to constrained funds and capacities (Dodman et al., 2022). At the beginning of the pandemic, and as it continued, a decline in cities' tax base translated into limited local recovery spending (Kunzmann, 2020). This, in turn, led to a requirement for emergency liquidity to relieve debt payments and increase borrowing capacity to expand and upgrade healthcare capacity, fund testing and tracing, maintain the operation of transport systems while implementing social distancing, and support small businesses affected by lockdown measures, among other issues. The capacity to promote climate-resilient development in cities has thus been compromised by the pandemic response, as the prioritization of investment has postponed actions and, in some cases, scaled back short-term ambitions, particularly in developing countries (Corfee-Morlot et al., 2021; ECLAC, 2022).

Spending during the COVID-19 pandemic revealed an egregious asymmetry not just between advanced and developing economies but also between immediate response (rescue) and recovery. Recovery spending clearly dominated in developed economies, while funding in developing countries was limited to rescue (O'Callaghan & Adam, 2021). Funding for mid- and long-term actions was – and still is – limited or absent in cities in developing economies; they are still struggling with the socioeconomic impacts of the pandemic. However, in both cases, there are certainly still few long-term commitments, either for expanding water and sanitation infrastructure and promoting renewable energy (including household-level solutions), or for pursuing urban retrofit and redesign (including public spaces and nature-based solutions), and even for programs aimed at increasing the flow of investment-ready projects that address climate-resilient development locally. The latter is a major issue as cities need to develop clear mechanisms for incorporating climate, environmental, and health goals into long-lasting infrastructure and land-use decisions.

Since developing and emerging market economies have been hit hardest by both climate change and COVID-19, unlocking new sources of funding for sustainable recovery in those countries – which will be paid back many times over (Gulati et al., 2020) – is particularly urgent. Among funding mechanisms are green bonds – as used in the case of Transport for London (TfL, 2020) – and international sustainability funds. The latter can be linked to SDGs to locally "build back better" from the COVID-19 crisis by, for example, prioritizing investments through robust green budgeting tools that increase the efficiency, effectiveness, and impacts of budgetary processes. For instance, priority funding can be designated for those projects that provide the most climate, environmental, and health cobenefits (C40, 2021; WHO, 2021a).

Increasing national–local collaboration and coordination of actions can enable more ambitious nationally determined contributions (NDCs) in which cities can play a central role (Gulati et al., 2020).[19] To support urban transformative action, which will demand an estimated 93 trillion US dollars worth of sustainable infrastructure by 2030 (NCE, 2016), funding partnerships and initiatives have recently been launched such as the City Resilience Program (CRP) and the City Climate Finance Gap Fund. Implemented by the World Bank and European Investment Bank, the City Climate Finance Gap Fund aims to help bridge the urban financing gap to achieve low carbon, climate-resilient urbanization pathways. The CRP is a partnership between the World Bank and the Global Facility for Disaster Reduction and Recovery (GFDRR), which is anchored in the former. Launched in June 2017 as a multi-donor initiative aimed at increasing financing for urban resilience, the CRP is supported by the Swiss State Secretariat for Economic Affairs (SECO) and the Austrian Federal Ministry of Finance.

In its first year of operation, the City Climate Finance Gap Fund supported thirty-three cities worldwide. Funding for disadvantaged communities and

---

[19] See ARC3.3 Element on *Financing Climate Action*, www.cambridge.org/core/publications/elements/elements-in-climate-change-and-cities.

informal urban areas is central to closing current divides and advancing just climate-resilient development (ECLAC, 2021; IPCC, 2022c). Investments therefore need to prioritize actions to restore and improve urban services, promote nonconventional forms of renewable energy, advance circular economy schemes, restore local and peri-urban ecosystems, encourage digital economy and sustainable tourism, and strengthen healthcare manufacturing industries and the healthcare economy (ECLAC, 2021). In the case of developing countries, the Economic Commission for Latin America and the Caribbean (ECLAC) recommends that investments should additionally seek to transform the prevailing productive structure to create jobs, stimulate the economy, reduce imports and therefore indebtedness, as well as to reduce carbon footprint and other environmental impacts (ECLAC, 2021). (See the additional resources for a further discussion of COVID-19 and urban finance.)

7.1 Energy Consumption, Renewables, and Trends

The impacts of the COVID-19 pandemic on the global economy and its performance have significantly changed global energy use and energy supply chain structures. An International Energy Agency (IEA) study estimated that total primary energy demand decreased by about 6 percent globally in 2020; this reduction was approximately seven times greater than during the global financial crisis of 2008–2009. Furthermore, approximately 8 percent of the forty million jobs in the energy sector were either at risk or lost in 2020 (IEA, 2020). In this context, all renewable energy producers, apart from China, decreased the volume of their energy supply (Bhuiyan et al., 2021). However, during lockdowns the power mix shifted toward renewables (IEA, 2021), following the reduction in electricity demand (in China, for example, coal-fired power generation decreased). With lockdown relaxation, the mix nonetheless went back to pre-COVID-19 trends with further increases in the volume of energy demanded in countries such as India (IEA, 2021).

The USA is among the few countries that did not register a significant reduction in total energy demand: Increases in residential energy consumption (there was an increase of about 10 percent during the second quarter of 2020 in relation to the average registered between 2016 and 2019) practically offset the decline in business and industrial demand, which has been estimated to be about 12 percent and 14 percent, respectively (Cicala, 2020). This was very likely due to people spending more time at home (Figure 9).

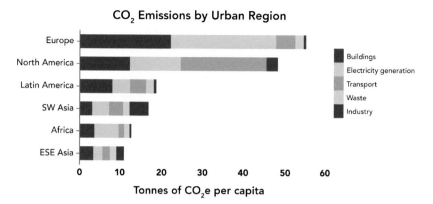

Figure 9 Sources of CO_2 emissions by urban region (C40, 2021).

Despite the general reduction in energy consumption during lockdowns, a noticeable increase in energy consumption at the household level was reported, not only due to greater demand for air conditioning (heating and cooling) and cooking but also to an increase in the use of domestic electrical and electronics equipment (Cuerdo-Vilches et al., 2021; Surahman et al., 2022). Households used cooling/heating devices for longer durations during the period when they were confined at home, believing that energy savings could cause thermal discomfort (as documented in half of the households in Madrid [Cuerdo-Vilches et al., 2021]). Energy consumption peaks related to cooling and heating at the household level were reported during the day and not just in the evenings, as observed before the COVID-19 pandemic in Quebec City (Rouleau & Gosselin, 2021).

Understanding such changes in household energy consumption can help anticipate shocks to electricity systems during potential future crises. Tailored urban energy strategies are needed to ensure energy justice while advancing urban sustainability. Despite the opportunity that the COVID-19 pandemic presented to improve residential energy efficiency and decarbonize, it should be expected that the population will revert back to pre-COVID behavior with no permanent gains in either. Instead of bouncing back to pre-COVID-19 business-as-usual practices, transitioning toward low-carbon and sustainable urban living is needed to advance energy and material efficiency, both in general and especially for buildings. This needs to go hand in hand with energy transition policies that enable increased local generation of renewable energy (León & Cárdenas, 2020; PNUMA, 2021) along with energy-efficient building design, construction, and operations (IPCC, 2022d).

7.2 Urban Supply Chain Disruptions

Many urban supply chains were significantly affected in the acute phase of the COVID-19 pandemic. For example, food-supply chains were significantly disrupted during the pandemic and the resilience of the most vulnerable urban populations suffered because of transport interruptions, labor shortages, consumer panic-purchasing, and changes in food consumption patterns (Batty, 2020; Newell & Dale, 2020; Pulighe & Lupia, 2020). To overcome these situations, local food production that supports small and medium-sized enterprises or locally based cooperatives could enhance local food-value chains (Carducci et al., 2021) (see Case Study 2). However, local food production may not be able to significantly reduce.

Linking local food production schemes to energy generation and/or nutrient recovery from food production residues – and, in general, from municipal organic waste – can support urban circularity. In small urban settlements surrounded by agricultural activities, there is a clear opportunity to profit. The case of Feliz municipality in Brazil demonstrates the power generation potential, equivalent to 10 percent of the total energy consumption of the municipality (PNUMA, 2021). (For further details, see the additional resources.)

7.3 Economic Impacts and Implications for Cities

The COVID-19 pandemic affected the global economy (ILO, 2020, 2021, 2023; IMF, 2021). Lowest recoveries have been experienced by low-income countries, as expressed in terms of GDP per capita. As reported by the International Labour Organization (ILO), labor markets have deteriorated since 2020 with decent work deficits persisting around the world and highly unequal labor market prospects across and within countries. Informal work has increased, productivity growth has slowed, and incentives for productive investments have weakened. In this context, trade and tourism have been particularly impacted: a decrease of 5.4 percent in the volume of international trade was observed during 2020 despite the economic and fiscal measures taken by governments (CEPAL, 2021; PNUMA, 2021).

For some cities whose economy mostly relies on tourism activities, this meant higher losses and harder recovery slopes. This has been documented in the case of tourist destinations in Latin American and the Caribbean (CEPAL, 2020), South Africa (Rogerson & Rogerson, 2020), some Chinese destinations (Qiu et al., 2020), and Nepal (Sah et al., 2020). Evidence of such impacts has led to the recognition of the importance of diversifying urban economies and advancing circular economies (UNEP & UN-Habitat, 2021; UN-Habitat, 2022). Urban resilience building has been identified as a risk-reducing factor against the impacts of a global crisis such as the COVID-19 pandemic as well as climate change.

The digitalization of work activities not only devolved part of the operational costs to employees (Battisti et al., 2022); it aslo changed the housing market in several cities (see Section 6.3). Returning to face-to-face work activities has encountered resistance, particularly in US cities. At the beginning of 2023, office occupancy in ten major US cities exceeded 50 percent for the first time since the pandemic started. As of February 27, 2023, the city with the highest weekly occupancy was Houston with 60.6 percent, while San Jose exhibited the lowest weekly occupancy at 41.2 percent (Kastle, 2023). Macroeconomic slow-down at the national level could have consequences for both a country and a city's future ability to borrow and invest in low-carbon economies, particularly in the absence of actions that seek to transcend business-as-usual practices (Corfee-Morlot et al., 2021; Hepburn et al., 2020; Watkins et al., 2021).

However, further assessments are needed for a better understanding of COVID-19 economic impacts at the urban level, particularly on the future of urban adaptation and mitigation agendas. This last point is important as, across a range of cases, pre-pandemic budgets for national and local climate action programs were partially redirected to COVID-19 response (Delgado Ramos, 2023; IMF, 2021). Cities' ability to meet their climate ambitions is thus considered to be at a critical juncture (Cities Climate Finance Leadership Alliance et al., 2021), despite recognition that they need to be prepared for dynamic and unpredictable futures (UN-Habitat, 2022).

8 Learning from COVID-19: Accelerating Urban Transformational Pathways

The word is experiencing simultaneous crises of conflict, inequity, and poverty. At the same time, the AR6 Summary for Policymakers (IPCC, 2022d) argues that the time to respond to climate change is "now or never." This Learning from COVID-19 Element reiterates the need to identify synergies between climate action and multiple ongoing crises. It also flags the need to intervene when the ongoing agenda of building climate resilience at the city level is derailed by large-scale disasters, such as the COVID-19 pandemic. The consequences of the pandemic will continue to affect the structure of urban systems for years to come.

8.1 Enabling Conditions

Urban planning and design and infrastructure development are key enablers of synergistic actions to address climate change adaptation and mitigation as well as COVID-19 or any future pandemics.[20] Emissions reduction and health considerations cannot be addressed through behavioral changes alone but require urban planning, infrastructure provisioning, and technological change. Investments in urban planning, infrastructure, and technology will provide opportunities to revitalize urban economies, which went into decline during the pandemic.

Acceleration of energy efficiency retrofitting, fuel shifting, and net-zero construction will reduce urban emissions and generally improve public health, particularly if future pandemics require people to spend more time at home (Lwasa & Seto, 2022). Enhancing urban investment and diversifying cities' revenue sources will facilitate coordinated and intersectional actions to address the simultaneous challenges of poverty and inequity, climate change, and public health.[21]

Social capital can provide several benefits during crisis scenarios, and high social capital communities respond more efficiently than those with low social capital. This is an important enabling condition to cope with pandemics and enhance community-level resilience to climate change through undertaking local-level adaptation and mitigation actions. Partnerships between various actors and institutions at local level – among, for example, community-based organizations (CBOs) and NGOs; CBOs and NGOs with local government and the private sector; and local government with the private sector – could be observed when managing COVID-19.

These partnerships are also important enabling conditions for local climate change actions. Institutional arrangements will be needed to enhance the science–policy interface as well as for setting up and nurturing local organizations (Solecki et al., 2021; UNEP & UN-Habitat, 2021). At the same time, these partnerships require financial support, not just from public funds but also through philanthropic and corporate social responsibility (CSR) initiatives, all within a context of recognizing local realities and priorities.

The barriers to dual action on climate change mitigation/adaptation and managing pandemics increase when poverty and inequity prevail. These are structural barriers to global cooperation such as material endowments, political systems, and ideas, values, and belief systems (Dubash et al., 2022). The COVID-19

---

[20] See ARC3.3 Element on *Urban Planning, Design, and Architecture*, www.cambridge.org/core/publications/elements/elements-in-climate-change-and-cities.

[21] See ARC3.3 Element on *Financing Climate Action*, www.cambridge.org/core/publications/elements/elements-in-climate-change-and-cities.

response in many countries led to a backsliding of democratic decision-making processes and increased centralization of directives at national government level. There is fear of such a governance approach becoming permanent, thereby moving away from the goal of integrated governance, equity, and sustainable development (Dubash et al., 2022).

8.2 Conclusions

Urgent responses to the COVID-19 pandemic spanned multiple sectors and scales, and were prioritized over responses to climate change. The immediate and significant health crisis resulted in governments at local, regional, national, and international levels diverting resources that could have been utilized for climate change mitigation and adaptation. In many countries, the early response to the pandemic was for national governments to take decision-making away from local entities, consequently reinforcing centralization practices. Often, governments drew on the expertise of civil society and industry to combat the pandemic, with nongovernmental organizations and local governments partnering to support informal sector residents and low-income populations. The pandemic also resulted in displays of urgency in dealing with the crises, with cities coordinating actions and incorporating ICTs to generate and disseminate information, all of which can be applied to planning for climate change mitigation and adaptation moving forward.

City size, form, and granularity had mixed impacts on COVID-19 transmission, but these are important for climate change mitigation strategies. The relationship between density and the spread of COVID-19 infections has been found to be relatively weak, contrary to initial thinking that a high correlation would be found between densely built or populated cities and the rapid spread of infectious diseases. The relationship between density and urban health is complex and mediated by the interplay between exposure, sensitivity, and the adaptive capacities of cities which extend into issues of urban inequalities. Crowding is responsible for the spread of infection. If urban density had been found to be highly correlated with the spread of COVID-19, then it would have presented further challenges to local governments who had been working to increase urban density as an important intervention in climate change mitigation actions. Pandemics such as COVID-19 can be managed, even in high-density areas, provided that residents make changes in lifestyle and follow public health instructions. Thus, policymakers are encouraged to promote compact cities, which are beneficial for climate change mitigation, even when handling pandemics such as COVID-19.

Climate change-related disasters and COVID-19 present a dual health threat to cities. Climate change-related disasters during the COVID-19 pandemic created a double health crisis. During flooding, heat waves, hurricanes, and other climate-related disasters, the emergency responses required may conflict with the emergency response and relief practices needed during a pandemic. Thus, healthcare personnel may be diverted to attend to disaster-related health concerns during an ongoing pandemic. Relief centers operationalized during climate-induced disasters need to be planned and operated in such a way as to make these pandemic-proof, that is, to minimize disease spread.

The COVID-19 pandemic has shifted populations from urbanized to less dense areas resulting in greater use of household energy, materials, and land. There is a real danger of reversal of the "back to the city movement" because of the COVID-19 pandemic. This will undo the gains of the urban redevelopment and regeneration and transit-oriented development projects that have been implemented to combat sprawl as part of climate change mitigation in countries around the world. The fear of infection as well as measures aimed at mitigating it have caused people to avoid public transit and vertical transportation systems such as elevators to minimize close contact with others, thus leading to a desire to move to single-family dwellings in the suburbs. Out-migration to peri-urban areas is further reinforced by the possibility of teleworking. Remote working brings with it a preference for larger homes with adequate workspaces.

The COVID-19 pandemic caused a transport modal shift that conflicts with climate change mitigation. The COVID-19 pandemic has led to decreased use of public transportation and other forms of shared mobility and increased preference for individual mobility, commonly typified by the personal automobile. During the pandemic, cities continued to support residents to take public transport by implementing measures related to safety. However, these measures have the potential to increase the cost of public transportation, which then puts an additional burden on local or national governments. The trend of prioritizing private over public transport may continue after the pandemic and set back the gains of sustainable and low-carbon transportation pathways that many cities have developed and promoted.

Built environment changes instigated as COVID-19 responses can also be climate change mitigation and adaptation actions. The pandemic has led to cities changing the use of their built environments, such as repurposing sidewalks for the extension of cafés to achieve the required social distancing and converting parking spaces into public open spaces. These changes are consonant with built environment changes that have been implemented for the purposes of climate change mitigation and adaptation. Many cities have encouraged active transportation by replacing car lanes with bike and pedestrian lanes, transforming parking lots into seating areas, and closing off sections of streets, or entire streets, for nonvehicle travel. While these efforts need to be maintained in support of active transportation and climate change mitigation and adaptation efforts, a consensus has not yet been reached on whether these changes are continuing in the period of normalization of activities in cities now that COVID-19 has become endemic. At the level of individual buildings, measures to increase natural light and ventilation to create healthier indoor environments and reduce cooling demand should be prioritized due to the climate mitigation and pandemic management benefits.

The provision of urban green spaces – a proven climate change mitigation and adaptation strategy – and greening buildings resulted in reduced health risks during the COVID-19 pandemic. Access to urban green spaces has been associated with lower spread of COVID-19 in several cities throughout the world. Green spaces are important for carbon sequestration and hence climate change mitigation and also adaptation to climate change impacts such as urban heat islands, excess rains, and storms. City greening measures include not just increased access to green spaces, but also green roofs, ventilation corridors, and urban farming. Also, designing green spaces to support wildlife improves the resilience of both human and natural systems.

The COVID-19 pandemic put stress on supply chains. During the pandemic, government-mandated lockdowns, labor shortages, and changes in consumer purchasing behavior and locations disrupted supply chains and impacted community resilience to climate change. Shocks to the food-supply chain revealed the limited food access and storage capacity of communities, often resulting in external food aid. Strengthening local–regional production and supply chains can diversify the economy, create jobs, support urban circularity, and enable resilience at the core of urban planning and management.

The COVID-19 pandemic disproportionately affected people who are socially and economically disadvantaged and homeless, the same groups most impacted by climate change. Although the issue of inequity, particularly in urban areas, and how that plays a role in heightening vulnerability and risk exposure to natural disasters and other impacts of climate change is not new, the COVID-19 pandemic has magnified underlying socioeconomic inequalities. For instance, disadvantaged groups have disproportionately reduced access to basic services and utilities, thereby deepening poverty and health inequity by amplifying disparities in healthcare access and health outcomes. This highlights the need for greater action aimed at reducing these inequalities to improve urban resilience – as the pandemic has shown, nobody is safe until everyone is safe.

The COVID-19 pandemic engendered multiscalar and polycentric collaborative governance, partnerships, and sources of financial help. Governance collaborations were observed across different scales of government and, for metropolitan cities, across boundaries. Partnerships included civil society institutions and actors. The presence of social capital and trust among actors and institutions mattered. Funding for local actions was sourced from public, private, philanthropic, and sweat equity to respond to the emergency caused by the pandemic. Similar governance systems need to be put in place, sustainable in the long run and therefore institutionalized, for climate change actions.

City governments have a crucial role to play in the collaborative governance processes required to tackle COVID-19, improve the overall reduction in GHG emissions, and build cities' climate resilience. The COVID-19 pandemic (like other major crises) may have provided an opportunity to rethink the role of municipal governance and facilitate policy innovation and experimentation – a shift away from what is conceivable, feasible, and socially and politically acceptable. City governments have proven their ability to take action during a pandemic and they have been able to open up a window for sustained local-level engagement on health security issues and how to respond to future public health emergencies. Whether these innovations and initiatives can be sustained and lead cities toward a long-term transformative agenda remains to be seen.

8.3 Further Research

- This ARC3.3 Element reveals the need for more studies on city-specific details of how the pandemic impacted local government budgets, investment priorities, and future planning efforts related not only to development, but also to climate change mitigation and adaptation. It is also not yet known whether the multiscalar governance and stakeholder partnerships forged to address the health emergency have continued once the pandemic abated.
- More evidence is required on how COVID-19 and extreme heat were managed by informal settlements and low-income populations in cities of the Global South, to understand and be better prepared for future compound-vulnerability situations.
- Addressing the degree to which urban areas and their inhabitants, which carried most of the COVID-19 burden, overlapped with those affected by climate change impacts and environmental degradation may inform as to

where work should be prioritized to enable a more robust and integrated approach to building long-term alternative-solution pathways.
- During the COVID-19 pandemic, governance was highly centralized along with the deployment of invasive and intrusive digital tools. Their acceptance during the pandemic was on account of the emergency, but the question remains whether these would be good tools for decentralized and participatory urban governance in general and for climate actions.
- More understanding of social, class, and gender-based equity impacts at the city level is required as well as what these would mean for climate change mitigation efforts related to transport, housing, urban structure, and service provision.
- Further research on the effects of misinformation, lack of trust in science, and reductions in social capital is needed.

8.4 Final Thoughts

The way forward on climate change is to incorporate equity, fairness, and justice into context-specific actions related to the agendas on climate change and SDGs. The climate change actions need to be taken within the SDG framework, which enables the former to go hand in hand with the development agendas – the key message from the IPCC 1.5 Report (IPCC, 2018) and followed up in AR6, WGII, and WGIII reports (Dodman et al., 2022; Lwasa & Seto, 2022). While health is one of the SDGs, this Element draws attention to the COVID-19 pandemic and illustrates how such global episodic events can set back, or have the potential to push forward, climate actions. These relationships are not linear or ubiquitous and vary across the world depending on how closely national and local governments coordinate on agendas to meet the SDGs.

The COVID-19 pandemic raises alarms about future episodes of zoonoses on account of deforestation and biodiversity loss driven by increasing urbanization and expanding agricultural production. Countries in Asia and Africa are likely to experience future urbanization and may need to pay attention to this warning. The pandemic also provides lessons learned on the impacts of pandemics to daily urban life that can be reduced by multilevel and participatory governance, social protection measures, urban planning and design and infrastructural systems, and people's behavior. The examples presented in this Element also reveal how cities quickly adapted to the new situation by managing public transport; appropriating and repurposing streets; finding new ways to work, study, and seek healthcare; shift to low-carbon mobility; and forge multiple partnerships. However, while it is true that urban systems changed rapidly, many of those changes have been reversed and their longevity is not clear.

There are, however, *challenges*. These include reversing the movement from cities to suburbs, particularly in the sprawling cities of the Global North, formalizing informal sector housing, and providing resources at the local level for investments in mitigation technologies. These present resistance to achievement of the changes needed to create low-carbon, resilient, and equitable cities. The pandemic has been a *call to action* to reconnect science, policy, and participatory governance to implement solutions. It has provided an *opportunity* to reset the carbon-based economy through novel, yet contextualized, urban transformational pathways.

Appendix: UCCRN ARC3.3 Stakeholder Soundings

Discussion with City Stakeholders and UCCRN's South Asia Hub

The South Asia hub of UCCRN hosted a hybrid workshop on October 27, 2021 at Ahmedabad University, India. The hub facilitated discussions on the impacts of the COVID-19 pandemic on climate change efforts in cities and its effects on urban policy, financing resilience and low-carbon developments, and innovation. The group of sixty-six participants included climate experts, city officials, public health specialists, urban studies academics, and students.

Collaborating organizations included:

(1) The Global Centre for Environment and Energy (GCEE), Ahmedabad University
(2) International Urban and Regional Cooperation (IURC)
(3) The IPCC Ahmedabad Technical Support Unit.

Speakers included ARC3.3 authors, urban policymakers, and officials from the Climate Change Department of the Government of Gujarat. The case study cities represented included Boston, Texas, Bologna, Gangtok, Surat, Pune, Srinagar, Gwalior, and Jaipur.

Insights from Stakeholders

- Cities were already facing simultaneous challenges like the climate crisis, poverty and inequity, biodiversity loss, haphazard developments, overburdened infrastructure systems, environmental degradation, urban resilience, and emergency preparedness and the COVID-19 pandemic increased their burden.
- Cities perceive various challenges at the city level in a siloed manner, and yet perceive issues in an intersectional manner for accelerating more resilient, healthier, and sustainable urban transformation pathways.

- Although the COVID-19 pandemic and climate change are both global phenomenon, mitigation must be rooted in the local urban context.
- In India, urban local governments that are already resource constrained faced financial challenges in managing the spread of COVID-19, healthcare provisioning, and extending pandemic welfare.
- Slowed economic growth, exacerbated poverty, and unemployment may shift the attention to economic recovery instead of climate action in resource constrained regions like Indian cities.
- Access to local-level data is crucial in controlling the spread and severity of COVID-19. Rigorous and regular mapping can enhance local databases and support integrated planning.
- Command-and-control centers created to monitor the spread of COVID-19 and relief centers for the ill and quarantined can help manage future climate change-induced disasters.

Discussion with City Stakeholders and UCCRN's Mexico Hub

The knowledge platform for urban transformation, which has hosted UCCRN's Mexico hub since 2021, organized a virtual workshop on the urban climate change agenda during COVID-19 on November 23, 2021. It gathered together thirty-six federal and city officials and academics from fourteen cities in Mexico (Mexico City and its metropolitan area, Puebla, Cancun, Mexicali, Ciudad Juarez, Monterrey, and Saltillo), Brazil (Porto Alegre, Rio Grande do Sul, and São Paulo), Colombia (Medellin), Uruguay (Montevideo), Puerto Rico (San Juan), and Peru (Lima).

Insights from Stakeholders

- Appropriate responses to cope with COVID-19 and climate change impacts need to be locally contextualized as there is not a one-size-fits-all solution.
- With some exceptions, local capacities for facing COVID-19 impacts in Latin American cities have shown to be limited – a reality also shared in long-term urban environmental, climate change, and resilience governance at the local level. The COVID-19 pandemic has revealed the urgent need to advance local institutional capacity building, improve policymaking related to interinstitutional coordination at different spatial scales, and cultivate stronger collaborations with private and social sectors (social sector organizations are incorporated entities – nonprofits, for-profits, or hybrids of the two – whose main purpose is to achieve a social mission).
- In most cases, resources previously allocated to sustainability, climate change, and resilience were reoriented to confront the impacts of COVID-19 during the

first and second waves. Intervention plans for urban sustainability, climate change adaptation, and mitigation were therefore constrained and even abandoned in cities where these issues had not received much attention historically and where lockdown measures hit public finances the hardest. For example, Mexican border cities, such as Mexicali and Ciudad Juarez, received thousands of immigrants seeking entry into the USA, even though the border was closed.

- Informality was one of the greatest challenges in Latin American cities because of the effects of lockdown measures on the informal economy, lack of access to social security and public health services, limited access to public services (e.g., water) that are necessary for hygiene and human rights, and other reasons.
- Cities with economies dependent on tourism saw a general economic decline and sector-related loss of jobs, as observed in Cancun or Cartagena. Economic decline, lower wages, and unemployment deepened the vulnerability of the poorest at a time when climate change impacts were worsening.
- Women faced a double caretaking workload with roles in hospitals and care facilities as well as in the home. Women are also the most affected in terms of disproportionate domestic violence, unemployment, and rent transfers. The latter has been particularly noticeable in the case of domestic workers who were, for the most part, unemployed until vaccination was widely available. In this context, women's networks have emerged to provide mutual support or locally produce food, as in the cases of Mexico City, Mexico; Medellin, Colombia; and Porto Alegre, Brazil.
- The digital divide was a generalized barrier for controlling the spread of COVID-19 as well as for coping with the diverse impacts of the pandemic, not just in the labor market and education (through home office and virtual education) but even within the daily operational practice of governments.
- In the region, the COVID-19 pandemic offered the opportunity not only to reformulate key aspects of current urban economic structure and its implications on urban form and function, but also urban lifestyles and consumption patterns.

References

Achremowicz, H., & Kamińska-Sztark, K. (2020). Grassroots cooperation during the COVID-19 pandemic in Poland. *DisP – The Planning Review*, *56*(4), 88–97. https://doi.org/10.1080/02513625.2020.1906062.

Al Saidi, A. M. O., Nur, F. A., Al-Mandhari, A. S., El Rabbat, M., Hafeez, A., & Abubakar, A. (2020). Decisive leadership is a necessity in the COVID-19 response. *Lancet*, *396*(10247), 295–298. https://doi.org/10.1016/S0140-6736(20)31493-8.

Alayza, N., & Caldwell, M. (2021). Financing climate action and the COVID-19 Pandemic: An analysis of 17 developing countries. World Resources Institute Working Paper. www.wri.org/research/financing-climate-action-and-covid-19-pandemic.

Alhadedy, N. H., & Gabr, H. S. (2022). Home design features post-COVID-19. *Journal of Engineering and Applied Science*, *69*(1), Article 87. https://doi.org/10.1186/s44147-022-00142-z.

Aliyu Bununu, Y., & Bello, A. (2024). *The Impact of COVID-19 Lockdowns in Accra, Kumasi and Johannesburg: Linkages to Informal Settlements*. UCCRN Case Study Docking Station. https://bit.ly/3YmLlnk.

Allam, Z., Bibri, S. E., Jones, D. S., Chabaud, D., & Moreno, C. (2022). Unpacking the "15-Minute 1City" via 6 G, IoT, and digital twins: Towards a new narrative for increasing urban efficiency, resilience, and sustainability. *Sensors*, *22*(4), 1369. https://doi.org/10.3390/s22041369.

American Public Transportation Association. (2020). Cleaning and disinfecting transit vehicles and facilities during a contagious virus pandemic. White Paper. https://tinyurl.com/4uyhktej.

American Public Transportation Association. (2023). *Transit Workforce Shortage: Synthesis Report*. www.apta.com/wp-content/uploads/APTA-Workforce-Shortage-Synthesis-Report-03.2023.pdf.

Angel, S., & Blei, A. (2020). COVID-19 thrives in larger cities, not denser ones. SSRN Scholarly Paper No. 3672321. https://doi.org/10.2139/ssrn.3672321.

Arriola Vega, L. A., & Coraza de los Santos, E. (2020). *Immobile and Vulnerable: Migrants at Mexico's Southern Border at the Outset of Covid-19*. Report 08.14.20. Rice University's Baker Institute for Public Policy.

Atkinson, C. L. (2022). A review of telework in the COVID-19 pandemic: Lessons learned for work–life balance? *COVID*, *2*(10), 1405–1416. https://doi.org/10.3390/covid2100101.

Bai, X., Nagendra, H., Shi, P., & Liu, H. (2020). Cities: Build networks and share plans to emerge stronger from COVID-19. *Nature*, *584*(7822), 517–520. https://doi.org/10.1038/d41586-020-02459-2.

Basu, R., & Ferreira, J. (2021). Sustainable mobility in auto-dominated Metro Boston: Challenges and opportunities post-COVID-19. *Transport Policy*, *103*, 197–210. https://doi.org/10.1016/j.tranpol.2021.01.006.

Battisti, E., Alfiero, S., & Leonidou, E. (2022). Remote working and digital transformation during the COVID-19 pandemic: Economic–financial impacts and psychological drivers for employees. *Journal of Business Research*, *150*, 38–50. https://doi.org/10.1016/j.jbusres.2022.06.010.

Batty, M. (2020). The coronavirus crisis: What will the post-pandemic city look like? *Environment and Planning B: Urban Analytics and City Science*, *47*(4), 547–552. https://doi.org/10.1177/2399808320926912.

Baudier, P., Kondrateva, G., Ammi, C., Chang, V., & Schiavone, F. (2023). Digital transformation of healthcare during the COVID-19 pandemic: Patients' teleconsultation acceptance and trusting beliefs. *Technovation*, *120*, 102547. https://doi.org/10.1016/j.technovation.2022.102547.

Beck, M. J., & Hensher, D. A. (2020). Insights into the impact of COVID-19 on household travel and activities in Australia: The early days under restrictions. *Transport Policy*, *96*, 76–93. https://doi.org/10.1016/j.tranpol.2020.07.001.

Bel, J.-B., & Marengo, P. (2020). The impact of the COVID-19 pandemic on municipal waste management systems. ARC+ News, April 12. https://tinyurl.com/269tahw4.

Bele, M. Y., Sonwa, D. J., & Tiani, A. M. (2014). Local communities vulnerability to climate change and adaptation strategies in Bukavu in DR Congo. *Journal of Environment & Development*, *23*(3), 331–357. https://doi.org/10.1177/1070496514536395.

Benfer, E. A., Vlahov, D., Long, M. Y., Walker-Wells, E., Pottenger, J. L., Gonsalves, G., & Keene, D. E. (2021). Eviction, health inequity, and the spread of COVID-19: Housing policy as a primary pandemic mitigation strategy. *Journal of Urban Health*, *98*(1), 1–12. https://doi.org/10.1007/s11524-020-00502-1.

Berdejo-Espinola, V., Suárez-Castro, A. F., Amano, T., Fielding, K. S., Oh, R. R. Y., & Fuller, R. A. (2021). Urban green space use during a time of stress: A case study during the COVID-19 pandemic in Brisbane, Australia. *People and Nature*, *3*(3), 597–609. https://doi.org/10.1002/pan3.10218.

BEREC. (2021). *Study on Post Covid Measures to Close the Digital Divide: Final Report*. Body of European Regulators of Electronic Communications. https://tinyurl.com/2s4d7exa.

Berkowitz, R. L., Gao, X., Michaels, E. K., & Mujahid, M. S. (2020). Structurally vulnerable neighbourhood environments and racial/ethnic COVID-19 inequities. *Cities & Health*, *5*(sup 1), S59–S62. https://doi.org/10.1080/23748834.2020.1792069.

Bezgrebelna, M., McKenzie, K., Wells, S., Ravindran, A., Kral, M., Christensen, J., Stergiopoulos, V., Gaetz, S., & Kidd, S. A. (2021). Climate change, weather, housing precarity, and homelessness: A systematic review of reviews. *International Journal of Environmental Research and Public Health*, *18*(11), Article 11. https://doi.org/10.3390/ijerph18115812.

Bhide, A. (2021). Informal settlements, the emerging response to COVID and the imperative of transforming the narrative. *Journal of Social and Economic Development*, *23*(2), 280–289. https://doi.org/10.1007/s40847-020-00119-9.

Bhuiyan, M. A., An, J., Mikhaylov, A., Moiseev, N., & Danish, M. S. S. (2021). Renewable energy deployment and COVID-19 measures for sustainable development. *Sustainability*, *13*(8), Article 8. https://doi.org/10.3390/su13084418.

Bilbao, B., Steil L., Urbieta, I. R., Anderson, L., Pinto, C., Gonzalez, M. E., Millán, A., Falleiro, R. M., Morici, E., Ibarnegaray, V., Pérez-Salicrup, D. R., Pereira, J. M., & Moreno, J. M. (2020). Incendios forestales. In J. M. Moreno, C. Laguna-Defi, V. Barros, E. Calvo Buendía, J. A. Marengo, & U. Oswald Spring (Eds.), *Adaptación frente a los riesgos del cambio climático en los países iberoamericanos: Informe RIOCCADAPT* (pp. 459–524). McGraw-Hill.

Blanco, H. (2020). Implications of COVID-19 for urban planning. In G. C. Delgado Ramos & D. López (Eds.), *Las ciudades ante el COVID-19: Nuevas direcciones para la investigación urbana y las políticas públicas* (pp. 12–24). Plataforma de Conocimiento para la Transformación Urbana and INGSA. https://doi.org/10.5281/zenodo.3894075.

Bohle, D., Medve-Bálint, G., Šćepanović, V., & Toplišek, A. (2022). Riding the Covid waves: Authoritarian socio-economic responses of east central Europe's anti-liberal governments. *East European Politics*, *38*(4), 662–686. https://doi.org/10.1080/21599165.2022.2122044.

Botello Peñaloza, H. A., & Guerrero Rincón, I. (2022). Informal economy and income distribution in Ecuador during the COVID-19 pandemic. *Revista Facultad de Ciencias Económicas: Investigación y Reflexión*, *30*(1), 11–28. https://doi.org/10.18359/rfce.5206.

Bourdrel, T., Annesi-Maesano, I., Alahmad, B., Maesano, C. N., & Bind, M.-A. (2021). The impact of outdoor air pollution on COVID-19: A review of evidence from in vitro, animal, and human studies. *European Respiratory Review*, *30*(159), 200242. https://doi.org/10.1183/16000617.0242-2020.

Boyce, M. R., & Katz, R. (2021). COVID-19 and the proliferation of urban networks for health security. *Health Policy and Planning*, *36*(3), 357–359. https://doi.org/10.1093/heapol/czaa194.

Boza-Kiss, B., Pachauri, S., & Zimm, C. (2021). Deprivations and inequities in cities viewed through a pandemic lens. *Frontiers in Sustainable Cities*, *3*, 645914. https://doi.org/10.3389/frsc.2021.645914.

Bucsky, P. (2020). Modal share changes due to COVID-19: The case of Budapest. *Transportation Research Interdisciplinary Perspectives*, *8*, 100141. https://doi.org/10.1016/j.trip.2020.100141.

Buehler, R., & Pucher, J. (2022). Cycling through the COVID-19 pandemic to a more sustainable transport future: Evidence from case studies of 14 large bicycle-friendly cities in Europe and North America. *Sustainability*, *14*(12), 7293. https://doi.org/10.3390/su14127293.

Burnett, H., Olsen, J. R., Nicholls, N., & Mitchell, R. (2021). Change in time spent visiting and experiences of green space following restrictions on movement during the COVID-19 pandemic: A nationally representative cross-sectional study of UK adults. *BMJ Open*, *11*(3), e044067. https://doi.org/10.1136/bmjopen-2020-044067.

Buurman, J., Freiburghaus, M., & Castellet-Viciano, L. (2022). The impact of COVID-19 on urban water use: A review. *Water Supply*, *22*(10), 7590–7602. https://doi.org/10.2166/ws.2022.300.

C40. (2021). *Technical Report: The Case for a Green and Just Recovery*. C40 Cities. https://tinyurl.com/ykr9vbzf.

Cabrera-Cano, Á. A., Cruz, J. C. C. la, Gloria-Alvarado, A. B., Álamo-Hernández, U., & Riojas-Rodríguez, H. (2021). Asociación entre mortalidad por Covid-19 y contaminación atmosférica en ciudades mexicanas. *Salud Pública de México*, *63*(4), Article 4. https://doi.org/10.21149/12355.

Calafiore, A., Dunning, R., Nurse, A., & Singleton, A. (2022). The 20-minute city: An equity analysis of Liverpool City Region. *Transportation Research Part D: Transport and Environment*, *102*, 103111. https://doi.org/10.1016/j.trd.2021.103111.

Capolongo, S., Gola, M., Brambilla, A., Morganti, A., Mosca, E. I., & Barach, P. (2020). COVID-19 and healthcare facilities: A decalogue of design strategies for resilient hospitals. *Acta Biomedica Atenei Parmensis*, *91*(9-S), 50–60. https://doi.org/10.23750/abm.v91i9-S.10117.

Carducci, B., Keats, E. C., Ruel, M., Haddad, L., Osendarp, S. J. M., & Bhutta, Z. A. (2021). Food systems, diets and nutrition in the wake of COVID-19. *Nature Food*, *2*(2), Article 2. https://doi.org/10.1038/s43016-021-00233-9.

Cash, D., Adger, W. N., Berkes, F., Garden, P., Lebel, L., Olsson, P., Pritchard, L., & Young, O. (2006). Scale and cross-scale dynamics: Governance and information in a multilevel world. *Ecology and Society*, *11* (2), 8. https://doi.org/10.5751/ES-01759-110208.

Castro-Sánchez, L.E. (2024). Urban Mobility and Climate Change Mitigation: Local Actions Paradigms before and after COVID-19 Emergency. UCCRN Case Study Docking Station. https://bit.ly/3YmLlnk.

Cavada, M. (2023). Evaluate space after Covid-19: Smart city strategies for gamification. *International Journal of Human–Computer Interaction*, *39*(2), 319–330. https://doi.org/10.1080/10447318.2021.2012383.

CDC. (2022). *Requirement for Face Masks on Public Transportation Conveyances and at Transportation Hubs*.https://stacks.cdc.gov/view/cdc/116429.

CEPAL. (2020). *El COVID-19 y la crisis socioeconómica en América Latina y el Caribe* (No. 132). Comisión Económica para América Latina y el Caribe. www.cepal.org/sites/default/files/publication/files/46838/RVE132_es.pdf.

CEPAL. (2021). *Estudio Económico de América Latina y el Caribe: Dinámica laboral y politicas de empleo para una recuperación sostenible e inclusiva más allá de la crisis del COVID-19*. Comisión Económica para América Latina y el Caribe. https://bit.ly/4cWdWV7.

Champlin, C., Sirenko, M., & Comes, T. (2023). Measuring social resilience in cities: An exploratory spatio-temporal analysis of activity routines in urban spaces during Covid-19. *Cities*, 135, 10422. https://doi.org/10.1016/j.cities.2023.104220.

Chen, J. T., Waterman, P. D., & Krieger, N. (2020). COVID-19 and the unequal surge in mortality rates in Massachusetts, by city/town and zip code measures of poverty, household crowding, race/ethnicity, and racialized economic segregation. Howard Center for Population and Development Studies Working Paper Vol. 19, No. 2. https://bit.ly/3WjuH5C.

Cheshire, P., Hilber, C., & Schöni, O. (2021). *The Pandemic and the Housing Market: A British Story*. Centre for Economic Performance, London School of Economics and Political Science.

Chu, Z., Cheng, M., & Song, M. (2021). What determines urban resilience against COVID-19: City size or governance capacity? *Sustainable Cities and Society*, 75, 103304. https://doi.org/10.1016/j.scs.2021.103304.

Cicala, S. (2020). Powering work from home. NBER Working Paper No. 27937. National Bureau of Economic Research. https://doi.org/10.3386/w27937.

Cicala, S., Holland, S. P., Mansur, E. T., Muller, N. Z., & Yates, A. J. (2020). Expected health effects of reduced air pollution from COVID-19 social distancing. EBER Working Paper No. 27135. National Bureau of Economic Research. https://doi.org/10.3386/w27135.

Cities Alliance. (2021). *COVID-19 and Informality: Good Practices for Reducing Risk and Enhacing Resilience*. https://bit.ly/3LaDgdT.

Cities Climate Finance Leadership Alliance, Climate Policy Initiative, The World Bank, & Atlantic Council. (2021). *The State of Cities Climate Finance: Executive Summary*. https://bit.ly/4c4EkLT.

City of Boston. (2022). Results of digital equity assessment announced. www.boston.gov/news/results-digital-equity-assessment-announced.

City of Helsinki. (2020). Helsinki sets its sights on innovative green recovery – "We can offer projects that benefit the whole of Finland."

City of Los Angeles. (2020). *SmartLA 2028: Technology for a better Los Angeles*. https://bit.ly/4csfaqp.

City of Portland. (2020). *Portland Economic Relief and Stabilization Strategy*. https://bit.ly/3VLbjiC.

Clement, J., Esposito, G., & Crutzen, N. (2023). Municipal pathways in response to COVID-19: A strategic management perspective on local public administration resilience. *Administration & Society*, *55*(1), 3–29. https://doi.org/10.1177/00953997221100382.

Coccia, M. (2020). *Two Mechanisms for Accelerated Diffusion of COVID-19 Outbreaks in Regions with High Intensity of Population and Polluting Industrialization: The Air Pollution-to-Human and Human-to-Human Transmission Dynamics*. www.medrxiv.org/content/10.1101/2020.04.06.20055657v1.full.pdf.

Coelho, K., Mahadevia, D., & Williams, G. (2020). Outsiders in the periphery: Studies of the peripheralisation of low income housing in Ahmedabad and Chennai, India. *International Journal of Housing Policy*, *22*(4), 543–569. https://doi.org/10.1080/19491247.2020.1785660.

Cohen-Shacham, E., Walters, G., Janzen, C., & Maginnis, S. (Eds.). (2016). *Nature-Based Solutions to Address Global Societal Challenges*. International Union for Conservation of Nature. https://doi.org/10.2305/IUCN.CH.2016.13.en.

Cole, M. A., Ozgen, C., & Strobl, E. (2020). Air pollution exposure and Covid-19. Discussion Papers No. 20–13. Department of Economics, University of Birmingham. https://ideas.repec.org/p/bir/birmec/20-13.html.

Combs, T. S., & Pardo, C. F. (2021). Shifting streets COVID-19 mobility data: Findings from a global dataset and a research agenda for transport planning and policy. *Transportation Research Interdisciplinary Perspectives*, *9*, 100322. https://doi.org/10.1016/j.trip.2021.100322.

Cong, W. (2021). From pandemic control to data-driven governance: The case of China's health code. *Frontiers in Political Science*, *3*. www.frontiersin.org/articles/10.3389/fpos.2021.627959.

Corfee-Morlot, J., Depledge, J., & Winkler, H. (2021). COVID-19 recovery and climate policy. *Climate Policy*, *21*(10), 1249–1256. https://doi.org/10.1080/14693062.2021.2001148.

Corrêa, H. L., & Corrêa, D. G. (2021). The Covid-19 pandemic – Opportunities for circular economy practices among sewing professionals in the city of Curitiba-Brazil. *Frontiers in Sustainability*, *2*. www.frontiersin.org/articles/10.3389/frsus.2021.644309.

Cortis, N., & Blaxland, M. (2020). *Australia's Community Sector and COVID-19: Supporting Communities through the Crisis*. Australian Council of Social Service. https://bit.ly/4fiBDIO.

CRED & UNDRR. (2021). *2020: The Non-COVID Year in Disasters*. www.undrr.org/media/49057/download?startDownload=20240716.

Cuerdo-Vilches, T., Navas-Martín, M. Á., & Oteiza, I. (2021). Behavior patterns, energy consumption and comfort during COVID-19 lockdown related to home features, socioeconomic factors and energy poverty in Madrid. *Sustainability*, *13*(11), Article 11. https://doi.org/10.3390/su13115949.

Czödörová, R., Dočkalik, M., & Gnap, J. (2021). Impact of COVID-19 on bus and urban public transport in SR. *Transportation Research Procedia*, *55*, 418–425. https://doi.org/10.1016/j.trpro.2021.07.005.

D'Amelio, J. (2020). "Victory Gardens" for the war against COVID-19. *CBS News*, April 5. www.cbsnews.com/news/victory-gardens-for-the-war-against-covid-19/.

Daanen, H., Bose-O'Reilly, S., Brearley, M., Flouris, D. A., Gerrett, N. M., Huynen, M., Jones, H. M., Lee, J. K. W., Morris, N., Norton, I., Nybo, L., Oppermann, E., Shumake-Guillemot, J., & Van den Hazel, P. (2021). COVID-19 and thermoregulation-related problems: Practical recommendations. *Temperature: Multidisciplinary Biomedical Journal*, *8*(1), 1–11. https://doi.org/10.1080/23328940.2020.1790971.

Dablanc, L. (2023). Urban logistics and COVID-19. In J. Zhang & Y. Hayashi (Eds.), *Transportation amid Pandemics* (pp. 131–141). Elsevier. https://doi.org/10.1016/B978-0-323-99770-6.00002-8.

Dalglish, S. L. (2020). COVID-19 gives the lie to global health expertise. *The Lancet*, *395*(10231), 1189. https://doi.org/10.1016/S0140-6736(20)30739-X.

Damasceno, R. M., Cicerelli, R. E., Almeida, T. D., & Requia, W. J. (2023). Air pollution and COVID-19 mortality in Brazil. *Atmosphere*, *14*(1), 5. https://doi.org/10.3390/atmos14010005.

Das, S., Boruha, A., Banerjee, A., Raoniar, R., Nama, S., & Maurya, A. K. (2021). Impact of COVID-19: A radical modal shift from public to private transport mode. *Transport Policy*, *109*, 1–11. https://doi.org/10.1016/j.tranpol.2021.05.005.

Davis, S. J., Liu, Z., Deng, Z., Zhu, B., Ke, P., Sun, T., Guo, R., Hong, C., Zheng, B., Wang, Y., Boucher, O., Gentine, P., & Ciais, P. (2022). Emissions rebound from the COVID-19 pandemic. *Nature Climate Change*, *12*(5), Article 5. https://doi.org/10.1038/s41558-022-01332-6.

de Haas, M., Faber, R., & Hamersma, M. (2020). How COVID-19 and the Dutch "intelligent lockdown" change activities, work and travel behaviour: Evidence from longitudinal data in the Netherlands. *Transportation Research Interdisciplinary Perspectives*, *6*, 100150. https://doi.org/10.1016/j.trip.2020.100150.

De Vos, J. (2020). The effect of COVID-19 and subsequent social distancing on travel behavior. *Transportation Research Interdisciplinary Perspectives*, *5*, 100121. https://doi.org/10.1016/j.trip.2020.100121.

Delgado Ramos, G. C. (2021). Climate-environmental governance in the Mexico Valley Metropolitan Area: Assessing local institutional capacities in the face of current and future urban metabolic dynamics. *World*, *2*(1), Article 1. https://doi.org/10.3390/world2010003.

Delgado Ramos, G. C. (2023). Transformación urbana en tiempos de pandemia y postpandemia: Capacidades institucionales para la acción climática-ambiental y de resiliencia en Ciudad de México y Juárez. In M. Suárez Lastra, & A. Ziccardi Contigiani (Eds.), *Ciudades mexicanas y condiciones de habitabilidad en tiempos de pandemia* (pp. 401–460). Coordinación de Humanidades, UNAM, Mexico City. https://decadacovid.humanidades.unam.mx/wp-content/uploads/DCM_tomo-12_CA.pdf.

Delgado Ramos, G.C. (2024). *Atepetl Urban Agriculture Program for Mexico City* [Photograph]. Personal collection.

Delgado Ramos, G. C. (Ed.). (2024). *Evaluación de la política ambiental de la Ciudad de México, 2018–2022*. Consejo de Evaluación de la Ciudad de México, Government of Mexico City. https://bit.ly/3LkYCVI.

Delgado Ramos, G. C., & López, D. (Eds.). (2020). *Las ciudades ante el COVID-19: Nuevas direcciones para la investigación urbana y las políticas públicas*. Plataforma de Conocimiento para la Transformación Urbana and INGSA. https://doi.org/10.5281/zenodo.3894075.

Delgado Ramos, G. C., & Mac Gregor Gaona, M. F. (2020). *Índice de Capacidades Institucionales Climáticas-Ambientales Locales, ICI–CLIMA 2019: El caso de la Zona Metropolitana del Valle de México* (1.0). Plataforma de Conocimiento para la Transformación Urbana. https://doi.org/10.5281/zenodo.3984235.

Delgado Ramos, G. C., Mac Gregor Gaona, M. F., Ortega León, R., & De Luca Zuria, A. (2019). *Hacia una agenda coordinada de acción climática–ambiental para la Zona Metropolitana del Valle de México*. Zenodo. https://doi.org/10.5281/ZENODO.3491502.

Desmaison, B., Jaime, K., Córdova, P., Alarcón, L., & Gallardo, L. (2022). Collective infrastructures of care: Ollas Comunes defying food insecurity during the COVID-19 pandemic. *Urbanisation*, 7(1), 46–65. https://doi.org/10.1177/24557471221110951.

DLR. (2020). A second DLR study on COVID-19 and mobility – public transport wanes in popularity, private transport gains in importance. DLR Portal, September 28. https://bit.ly/3VOkf6O.

Dodman, D. B., Hayward, B., Pelling, M., Castan Broto, V., Chow, W., Chu, E., Dawson, R., Khirfan, L., McPhearson, T., Prakash, A., Zheng, Y., & Ziervogel, G. (2022). Cities, settlements and key infrastructure. In *Climate Change 2022: Impacts, Adaptation, and Vulnerability. Contribution of Working Group II to the Sixth Assessment Report of the Intergovernmental Panel on Climate Change*. Cambridge University Press. www.ipcc.ch/report/sixth-assessment-report-working-group-ii/.

Dogar, A. A., Shah, I., Mahmood, T., Elahi, N., Alam, A., & Jadoon, U. G. (2022). Impact of Covid-19 on informal employment: A case study of women domestic workers in Khyber Pakhtunkhwa, Pakistan. *PLOS ONE*, 17(12), e0278710. https://doi.org/10.1371/journal.pone.0278710.

Dubash, N. K., Mitchell, C., Boasson, E. L., Borbor-Cordova, M. J., Fifita, S., Haites, E., Jaccard, M., Jotzo, F., Naidoo, S., Romero-Lankao, P., Shlapak, M., Shen, W., & Wu, L. (2022). National and sub-national policies and institutions. In *IPCC, 2022: Climate Change 2022: Mitigation of Climate Change. Contribution of Working Group III to the Sixth Assessment Report of the Intergovernmental Panel on Climate Change* (pp. 1355–1450). Cambridge University Press.

Duque Franco, I., Ortiz, C., Samper, J., & Millan, G. (2020). Mapping repertoires of collective action facing the COVID-19 pandemic in informal settlements in Latin American cities. *Environment and Urbanization*, 32(2), 523–546. https://doi.org/10.1177/0956247820944823.

Dzigbede, K., Gehl, S. B., & Willoughby, K. (2020). Disaster resiliency of U.S. local governments: Insights to strengthen local response and recovery from the COVID-19 pandemic. *Public Administration Review*, 80(4), 634–643. https://doi.org/10.1111/puar.13249

Dzisi, E. K. J., & Dei, O. A. (2020). Adherence to social distancing and wearing of masks within public transportation during the COVID 19 pandemic.

Transportation Research Interdisciplinary Perspectives, *7*, 100191. https://doi.org/10.1016/j.trip.2020.100191.

ECLAC. (2020). *The Social Challenge in Times of COVID-19: Special Report COVID-19* (No. 3). Economic Commission for Latin America and the Caribbean. https://doi.org/10.18356/9789210054676.

ECLAC. (2021). *Resilient Institutions for a Transformative Post-Pandemic Recovery in Latin America and the Caribbean: Inputs for Discussion*. Economic Commission for Latin America and the Caribbean. https://bit.ly/3ym1RJC.

ECLAC. (2022). *How to Finance Sustainable Development: Recovery from the Effects of COVID-19 in Latin America and the Caribbean*. Economic Commission for Latin America and the Caribbean. https://bit.ly/3YjuIZC.

Edelmann, N., & Millard, J. (2021). Telework development before, during and after COVID-19, and its relevance for organizational change in the public sector. *ICEGOV '21: Proceedings of the 14th International Conference on Theory and Practice of Electronic Governance*, October, 436–443. https://doi.org/10.1145/3494193.3494252.

Ellen MacArthur Foundation. (2020). *Circular Economy and the Covid-19 Recovery: How Policymakers Can Pave the Way to a Low-Carbon and Prosperous Future*. https://bit.ly/3KOOtk7.

Fahlberg, A., Martins, C., de Andrade, M., Costa, S., & Portela, J. (2023). The Impact of the pandemic on poor urban neighborhoods: A participatory action research study of a "favela" in Rio de Janeiro. *Socius*, *9*, 23780231221137140. https://doi.org/10.1177/23780231221137139.

Fahlberg, A., Vicino, T. J., Fernandes, R., & Potiguara, V. (2020). Confronting chronic shocks: Social resilience in Rio de Janeiro's poor neighborhoods. *Cities*, *99*, 102623. https://doi.org/10.1016/j.cities.2020.102623.

Fatemi, F., Ardalan, A., Aguirre, B., Mansouri, N., & Mohammadfam, I. (2017). Social vulnerability indicators in disasters: Findings from a systematic review. *International Journal of Disaster Risk Reduction*, *22*, 219–227. https://doi.org/10.1016/j.ijdrr.2016.09.006.

Fattorini, D., & Regoli, F. (2020). Role of the chronic air pollution levels in the Covid-19 outbreak risk in Italy. *Environmental Pollution*, *264*, 114732. https://doi.org/10.1016/j.envpol.2020.114732.

Feng, M., Ren, J., He, J., Chan, F. K. S., & Wu, C. (2022). Potency of the pandemic on air quality: An urban resilience perspective. *Science of the Total Environment*, *805*, 150248. https://doi.org/10.1016/j.scitotenv.2021.150248.

Field, C. B., Barros, V. R., Mach, K. J., Mastrandrea, M. D., van Aalst, M., Adger, W. N., et al. (2014). Technical summary. In *Climate Change 2014: Impacts, Adaptation, and Vulnerability. Part A: Global and Sectoral Aspects.*

Contribution of Working Group II to the Fifth Assessment Report of the Intergovernmental Panel on Climate Change (pp. 35–94). Cambridge University Press.

Finn, D. (2020). Streets, sidewalks and COVID-19: Reimaging New York City's public realm as a tool for crisis management. *Journal of Extreme Events*, *7*(4), 2150006. https://doi.org/10.1142/S2345737621500068.

Florida, R., & Gabe, T. (2023). COVID-19, the new urban crisis, and cities: How COVID-19 compounds the influence of economic segregation and inequality on metropolitan economic performance. *Economic Development Quarterly*, *37*(4), 328–348. https://doi.org/10.1177/08912424231155969.

Foster, S., Leichenko, R., Nguyen, K. H., Blake, R., Kunreuther, H., Madajewicz, M., Petkova, E., Zimmerman, R., Corbin-Mark, C., Yeampierre, E., Tovar, A., Herrera, C., & Ravenborg, D. (2019). New York City panel on climate change 2019 report chapter 6: Community-based assessments of adaptation and equity. *Annals of the New York Academy of Sciences*, *1439*, 126–173.

Fragkias, M., Lobo, J., Strumsky, D., & Seto, K. C. (2013). Does size matter? Scaling of CO_2 emissions and U.S. urban areas. *PLOS ONE*, *8*(6), e64727. https://doi.org/10.1371/journal.pone.0064727.

Frutos, P., Dominguez, D., Noguera, E., Parra, C., Montanía, C., & Cernuzzi, L. (2022). *The Role of ICTs in the Management of Citizen Initiatives Led by Women from Vulnerable Communities of Bañado Sur de Asunción during COVID-19*. *3321*. https://ceur-ws.org/Vol-3321/paper10.pdf.

Fundación Gonzalo Rodríguez & OPS. (2022). *COVID-19 y movilidad sostenible en América Latina: Los casos de Argentina, Chile y Uruguay*. Organización Panamericana de la Salud. https://iris.paho.org/bitstream/handle/10665.2/56012/ARG220002_spa.pdf?sequence=5.

Gao, J., & Zhang, P. (2021). China's public health policies in response to COVID-19: From an "authoritarian" perspective. *Frontiers in Public Health*, *9*. www.frontiersin.org/article/10.3389/fpubh.2021.756677.

Garaus, M., & Garaus, C. (2021). The impact of the Covid-19 pandemic on consumers' intention to use shared-mobility services in German cities. *Frontiers in Psychology*, *12*. https://www.frontiersin.org/article/10.3389/fpsyg.2021.646593.

Garavaglia, C., Sancino, A., & Trivellato, B. (2021). Italian mayors and the management of COVID-19: Adaptive leadership for organizing local governance. *Eurasian Geography and Economics*, *62*(1), 76–92. https://doi.org/10.1080/15387216.2020.1845222.

Geng, D. (Christina), Innes, J., Wu, W., & Wang, G. (2021). Impacts of COVID-19 pandemic on urban park visitation: A global analysis. *Journal* of Forestry Research, 32(2), 553–567. https://doi.org/10.1007/s11676-020-01249-w.

Gerard, F., Imbert, C., & Orkin, K. (2020). Social protection response to the COVID-19 crisis: Options for developing countries. *Oxford Review of Economic Policy*, *36*(Supp_1), S281–S296. https://doi.org/10.1093/oxrep/graa026.

Gil, D., Domínguez, P., Undurraga, E. A., & Valenzuela, E. (2021). The socioeconomic impact of COVID-19 in urban informal settlements. https://doi.org/10.1101/2021.01.16.21249935.

Gkiotsalitis, K., & Cats, O. (2021). Public transport planning adaption under the COVID-19 pandemic crisis: Literature review of research needs and directions. *Transport Reviews*, *41*(3), 374–392. https://doi.org/10.1080/01441647.2020.1857886.

Goutte, S., Péran, T., & Porcher, T. (2020). The role of economic structural factors in determining pandemic mortality rates: Evidence from the COVID-19 outbreak in France. *Research in International Business and Finance*, *54*, 101281. https://doi.org/10.1016/j.ribaf.2020.101281.

Guerin, T. F. (2021). Policies to minimise environmental and rebound effects from telework: A study for Australia. *Environmental Innovation and Societal Transitions*, *39*, 18–33. https://doi.org/10.1016/j.eist.2021.01.003.

Gulati, M., Becqué, R., Godfrey, N., Akhmouch, A., Cartwright, A., Eis, J., Huq, S., Jacobs, M., King, R., & Rode, P. (2020). *The Economic Case for Greening the Global Recovery through Cities. Seven priorities for National Governments*. Coalition for Urban Transitions. https://bit.ly/45wenT4.

Güneralp, B., Reba, M., Hales, B. U., Wentz, E. A., & Seto, K. C. (2020). Trends in urban land expansion, density, and land transitions from 1970 to 2010: A global synthesis. *Environmental Research Letters*, *15*(4), 044015. https://doi.org/10.1088/1748-9326/ab6669.

Guthold, R., Stevens, G. A., Riley, L. M., & Bull, F. C. (2018). Worldwide trends in insufficient physical activity from 2001 to 2016: A pooled analysis of 358 population-based surveys with 1.9 million participants. *The Lancet Global Health*, *6*(10), e1077–e1086. https://doi.org/10.1016/S2214-109X(18)30357-7.

Han, S., Sim, J., & Kwon, Y. (2021). Recognition changes of the concept of urban resilience: Moderating effects of COVID-19 pandemic. *Land*, *10*(10), Article 10. https://doi.org/10.3390/land10101099.

Harris, M. A., & Branion-Calles, M. (2021). Changes in commute mode attributed to COVID-19 risk in Canadian national survey data. *Findings*, February. https://doi.org/10.32866/001c.19088.

Hartman, T. K., Stocks, T. V. A., McKay, R., Gibson-Miller, J., Levita, L., Martinez, A. P., Mason, L., McBride, O., Murphy, J., Shevlin, M., Bennett, K. M., Hyland, P., Karatzias, T., Vallières, F., & Bentall, R. P. (2021). The authoritarian dynamic during the COVID-19 pandemic: Effects on nationalism and anti-immigrant sentiment. *Social Psychological and Personality Science*, *12*(7), 1274–1285. https://doi.org/10.1177/1948550620978023.

Hartmann, C., Hegel, C., & Boampong, O. (2022). The forgotten essential workers in the circular economy? Waste picker precarity and resilience amidst the COVID-19 pandemic. *Local Environment*, *27*(10–11), 1272–1286. https://doi.org/10.1080/13549839.2022.2040464.

Harvey, J. (2022). Covid-19's toll on the world's informal workers. *New Labor Forum*, *31*(1), 60–68. https://doi.org/10.1177/10957960211062873.

Hassankhani, M., Alidadi, M., Sharifi, A., & Azhdari, A. (2021). Smart city and crisis management: Lessons for the COVID-19 pandemic. *International Journal of Environmental Research and Public Health*, *18*(15), Article 15. https://doi.org/10.3390/ijerph18157736.

Heinrichs, H., Mueller, F., Rohfleisch, L., Schulz, V., Talbot, S. R., & Kiessling, F. (2022). Digitalization impacts the COVID-19 pandemic and the stringency of government measures. Scientific Reports, *12*, Article 21628. https://doi.org/10.1038/s41598-022-24726-0.

Heo, S., Desai, M. U., Lowe, S. R., & Bell, M. L. (2021). Impact of changed use of greenspace during COVID-19 pandemic on depression and anxiety. *International Journal of Environmental Research and Public Health*, *18*(11), Article 11. https://doi.org/10.3390/ijerph18115842.

Hepburn, C., O'Callaghan, B., Stern, N., Stiglitz, J., & Zenghelis, D. (2020). Will COVID-19 fiscal recovery packages accelerate or retard progress on climate change? *Oxford Review of Economic Policy*, *36*(S1), S359–S381. https://doi.org/10.1093/oxrep/graa015.

Hertwich, E., Lifset, R., Pauliuk, S., Heeren, N., Ali, S., Tu, Q., Ardente, F., Berrill, P., Fishman, T., Kanaoka, K., Kulczycka, J., Makov, T., Masanet, E., & Wolfram, P. (2020). *Resource Efficiency and Climate Change: Material Efficiency Strategies for a Low-Carbon Future*. IRP and UNEP. https://doi.org/10.5281/ZENODO.3542680.

Hicks, R. (2020). Covid-19 has hobbled Asia's recycling trade as demand for recycled plastic dips and recyclers face ruin. *Eco-Business*, August 30. https://bit.ly/3XvBjjd.

Hillis, S., N'konzi, J.-P. N., Msemburi, W., Cluver, L., Villaveces, A., Flaxman, S., & Unwin, H. J. T. (2022). Orphanhood and caregiver loss among children based on new global excess COVID-19 death estimates.

JAMA Pediatrics, *176*(11), 1145–1148. https://doi.org/10.1001/jamapediatrics.2022.3157.

Holbig, H. (2022). Navigating the dual dilemma between lives, rights and livelihoods: COVID-19 responses in China, Singapore, and South Korea. *Zeitschrift Für Vergleichende Politikwissenschaft*, *16*(4), 707–731. https://doi.org/10.1007/s12286-023-00555-x.

Holleran, M. (2022). Pandemics and geoarbitrage: Digital nomadism before and after COVID-19. *City*, *26*(5–6), 831–847. https://doi.org/10.1080/13604813.2022.2124713.

Hong, S., & Choi, S.-H. (2021). The urban characteristics of high economic resilient neighborhoods during the COVID-19 pandemic: A case of Suwon, South Korea. *Sustainability*, *13*(9), 4679. www.mdpi.com/2071-1050/13/9/4679.

Hörcher, D., Singh, R., & Graham, D. J. (2022). Social distancing in public transport: Mobilising new technologies for demand management under the Covid-19 crisis. *Transportation*, *49*(2), 735–764. https://doi.org/10.1007/s11116-021-10192-6.

Hourcade, J. C., Glemarec, Y., Coninck, H. de, Bayat-Renoux, F., Ramakrishna, K., & Revi, A. (2021). *Scaling up Climate Finance in the Context of COVID-19: A Science-Based Call for Financial Decision-Makers*. Green Climate Fund. https://bit.ly/3KOwXfE.

Hu, B., Guo, H., Zhou, P., & Shi, Z.-L. (2021). Characteristics of SARS-CoV-2 and COVID-19. *Nature Reviews Microbiology*, *19*(3), 141–154. https://doi.org/10.1038/s41579-020-00459-7.

IAPT. (2020). *Public Transport Authorities and COVID-19: Response from the Front Line*. International Association of Public Transport. https://bit.ly/3VESD34.

IDB. (2020). *Guía de vías emergentes para ciudades resilientes*. Banco Interamericano de Desarrollo. https://bit.ly/4cqDl8I.

IEA. (2020). *Sustainable Recovery: World Energy Outlook Special Report*. International Energy Agency. www.iea.org/reports/sustainable-recovery.

IEA. (2021). COVID-19 impact on electricity. Website. www.iea.org/reports/covid-19-impact-on-electricity.

IEA. (2022a). Transport. Website. www.iea.org/reports/transport.

IEA. (2022b). *Data Centres and Data Transmission Networks*. www.iea.org/energy-system/buildings/data-centres-and-data-transmission-networks.

ILO. (2020). *COVID-19 and the Tourism Sector*. International Labour Organization. https://digitallibrary.un.org/record/3862311.

ILO. (2021). *COVID-19 and the World of Work: Updated Estimates and Analysis*. International Labour Organization. www.ilo.org/media/7901/download.

ILO. (2023). *World Employment and Social Outlook: Trends 2023*. International Labour Organization. https://bit.ly/3xsesuA.

IMF. (2021). *World Economic Outlook: Managing Divergent Recoveries*. International Monetary Fund. www.imf.org/en/Publications/WEO/Issues/2021/03/23/world-economic-outlook-april-2021.

Indian Express. (2020). Gujarat unlock 4.0: Public parks to reopen, no time limit for shops. *Indian Express*, September 2. https://bit.ly/3XCU5pe.

Indorewala, H., & Wagh, S. (2020). How strong is the link between Mumbai's slums and the spread of the coronavirus? *Scroll.In*, July 20. https://bit.ly/3zcRhoA.

IPCC. (2018). *Global Warming of 1.5°C: An IPCC Special Report on the Impacts of Global Warming of 1.5°C above Pre-Industrial Levels and Related Global Greenhouse Gas Emission Pathways, in the Context of Strengthening the Global Response to the Threat of Climate Change, Sustainable Development, and Efforts to Eradicate Poverty*. www.ipcc.ch/sr15/.

IPCC. (2022a). Annex I: Glossary. In *Climate Change 2022: Mitigation of Climate Change – Contribution of Working Group III to the Sixth Assessment Report of the Intergovernmental Panel on Climate Change* (pp. 1793–1817). Cambridge University Press. www.ipcc.ch/report/ar6/wg3/downloads/report/IPCC_AR6_WGIII_FOD_AnnexI.pdf.

IPCC. (2022b). Annex II: Glossary. In *Climate Change 2022: Impacts, Adaptation and Vulnerability – Contribution of Working Group II to the Sixth Assessment Report of the Intergovernmental Panel on Climate Change* (pp. 2897–2930). Cambridge University Press. www.ipcc.ch/report/ar6/wg2/downloads/report/IPCC_AR6_WGII_Annex-II.pdf.

IPCC. (2022c). Energy systems. In *Climate Change 2022: Mitigation of Climate Change – Contribution of Working Group III to the Sixth Assessment Report of the Intergovernmental Panel on Climate Change*. Cambridge University Press. www.ipcc.ch/report/ar6/wg3/.

IPCC. (2022d). Summary for policymakers. In *Climate Change 2022: Mitigation of Climate Change – Contribution of Working Group III to the Sixth Assessment Report of the Intergovernmental Panel on Climate Change*. Cambridge University Press. www.ipcc.ch/report/ar6/wg3/.

Jaramillo, P., Ribeiro, S. K., Newman, P., Dhar, S., Diemuodeke, O. E., Kajino, T., Lee, D. S., Nugroho, S. B., Ou, X., Hammer Strømman, A., & Whitehead, J. (2022). Transport. In *Climate Change 2022: Mitigation of*

Climate Change – Contribution of Working Group III to the Sixth Assessment Report of the Intergovernmental Panel on Climate Change (pp. 1049–1144). Cambridge University Press. https://doi.org/10.1017/9781009157926.012.

Jiménez Cisneros, B. (2020). Nuevas direcciones para la investigación urbana y las política públicas en el ámbito del agua. In G. C. Delgado Ramos & D. López García (Eds.), *Cities and COVID-19: New Directions for Urban Research and Public Policies* (pp. 92–97). Plataforma de Conocimiento para la Transformación Urbana and INGSA.

Kabisch, N., Korn, H., Stadler, J., & Bonn, A. (Eds.). (2017). *Nature-Based Solutions to Climate Change Adaptation in Urban Areas: Linkages between Science, Policy and Practice*. Springer International Publishing. https://doi.org/10.1007/978-3-319-56091-5.

Kang, H., An, J., Kim, H., Ji, C., Hong, T., & Lee, S. (2021). Changes in energy consumption according to building use type under COVID-19 pandemic in South Korea. *Renewable and Sustainable Energy Reviews*, *148*, 111294. https://doi.org/10.1016/j.rser.2021.111294.

Kastle. (2023). Getting America back to work. Website. www.kastle.com/safety-wellness/getting-america-back-to-work/.

Khambule, I. (2022). COVID-19 and the informal economy in a small-town in South Africa: Governance implications in the post-COVID era. *Cogent Social Sciences*, *8*(1), 2078528. https://doi.org/10.1080/23311886.2022.2078528.

Khan, S., Mishra, J., Ahmed, N., Onyige, C. D., Lin, K. E., Siew, R., & Lim, B. H. (2022). Risk communication and community engagement during COVID-19. *International Journal of Disaster Risk Reduction*, *74*, 102903. https://doi.org/10.1016/j.ijdrr.2022.102903.

Khomenko, S., Cirach, M., Pereira-Barboza, E., Mueller, N., Barrera-Gómez, J., Rojas-Rueda, D., Hoogh, K. de, Hoek, G., & Nieuwenhuijsen, M. (2021). Health impacts of the new WHO air quality guidelines in European cities. *The Lancet Planetary Health*, *5*(11), e764. https://doi.org/10.1016/S2542-5196(21)00288-6.

Kidd, S. A., Greco, S., & McKenzie, K. (2021). Global climate implications for homelessness: A scoping review. *Journal of Urban Health*, *98*(3), 385–393. https://doi.org/10.1007/s11524-020-00483-1.

Knox-Hayes, J., Osorio, J. C., Stamler, N., Dombrov, M., Winer, R., Smith, M. H., Blake, R. A., & Rosenzweig, C. (2023). The compound risk of heat and COVID-19 in New York City: Riskscapes, physical and social factors, and interventions. *Local Environment*, *28*(6), 699–727. https://doi.org/10.1080/13549839.2023.2187362.

Ku, D.-G., Um, J.-S., Byon, Y.-J., Kim, J.-Y., & Lee, S.-J. (2021). Changes in passengers' travel behavior due to COVID-19. *Sustainability*, *13*(14), 7974. https://doi.org/10.3390/su13147974.

Kummitha, R. K. R. (2020). Smart technologies for fighting pandemics: The techno- and human- driven approaches in controlling the virus transmission. *Government Information Quarterly*, *37*(3), 101481. https://doi.org/10.1016/j.giq.2020.101481.

Kunzmann, K. R. (2020). Smart cities after COVID-19: Ten narratives. *DisP – The Planning Review*, *56*(2), 20–31. https://doi.org/10.1080/02513625.2020.1794120.

Kutralam-Muniasamy, G., Pérez-Guevara, F., Roy, P. D., Elizalde-Martínez, I., & Shruti, V. C. (2021). Impacts of the COVID-19 lockdown on air quality and its association with human mortality trends in megapolis Mexico City. *Air Quality, Atmosphere & Health*, *14*(4), 553–562. https://doi.org/10.1007/s11869-020-00960-1.

Lara-Valencia, F., & García-Pérez, H. (2021). The borders of the pandemic: Lessons on governance and cooperation in United States–Mexico border cities. *Estudios Fronterizos*, *22*, e067. https://doi.org/10.21670/ref.2104067.

Leal Filho, W., Salvia, A. L., Minhas, A., Paço, A., & Dias-Ferreira, C. (2021). The COVID-19 pandemic and single-use plastic waste in households: A preliminary study. *Science of the Total Environment*, *793*, 148571. https://doi.org/10.1016/j.scitotenv.2021.148571.

León, D. C., & Cárdenas, J. C. (2020). *Lessons from COVID-19 for a Sustainability Agenda in Latin America and the Caribbean*. UNDP Latin America and the Caribbean. https://bit.ly/4b6h8M4.

Li, F. (2022). Disconnected in a pandemic: COVID-19 outcomes and the digital divide in the United States. *Health & Place*, *77*, 102867. https://doi.org/10.1016/j.healthplace.2022.102867.

Li, S. L., Pereira, R. H. M., Prete Jr., C. A., Zarebski, A. E., Emanuel, L., Alves, P. J. H., Peixoto, P. S., Braga, C. K. V., de Souza Santos, A. A., de Souza, W. M., Barbosa, R. J., Buss, L. F., Mendrone, A., de Almeida-Neto, C., Ferreira, S. C., Salles, N. A., Marcilio, I., Wu, C.-H., Gouveia, N., Nascimento, V. H., Sabino, E. C., Fario, N. R., & Messina, J. P. (2021). Higher risk of death from COVID-19 in low-income and non-White populations of São Paulo, Brazil. *BMJ Global Health*, *6*(4), e004959. https://doi.org/10.1136/bmjgh-2021-004959.

Li, X., & Zhang, C. (2021). Did the COVID-19 pandemic crisis affect housing prices evenly in the U.S.? *Sustainability*, *13*(21), Article 21. https://doi.org/10.3390/su132112277.

Li, Y., Chandra, Y., & Fan, Y. (2022). Unpacking government social media messaging strategies during the COVID-19 pandemic in China. *Policy & Internet*, *14*(3), 651–672. https://doi.org/10.1002/poi3.282.

Lida, A., Yamazaki, T., Hino, K., & Yokohari, M. (2023). Urban agriculture in walkable neighborhoods bore fruit for health and food system resilience during the COVID-19 pandemic. *npj Urban Sustainability*, *3*, Article 4. https://doi.org/10.1038/s42949-023-00083-3.

Liu, J., Liu, S., Xu, X., & Zou, Q. (2022). Can digital transformation promote the rapid recovery of cities from the COVID-19 epidemic? An empirical analysis from Chinese cities. *International Journal of Environmental Research and Public Health*, *19*(6), 3567. https://doi.org/10.3390/ijerph19063567.

Liu, X., Huang, Y., Xu, X., Li, X., Li, X., Ciais, P., Lin, P., Gong, K., Ziegler, A. D., Chen, A., Gong, P., Chen, J., Hu, G., Chen, Y., Wang, S., Wu, Q., Huang, K., Estes, L., & Zeng, Z. (2020). High-spatiotemporal-resolution mapping of global urban change from 1985 to 2015. *Nature Sustainability*, *3*(7), Article 7. https://doi.org/10.1038/s41893-020-0521-x.

Liu, Z., Guo, J., Zhong, W., & Gui, T. (2021). Multi-level governance, policy coordination and subnational responses to COVID-19: Comparing China and the US. *Journal of Comparative Policy Analysis: Research and Practice*, *23*(2), 204–218. https://doi.org/10.1080/13876988.2021.1873703.

Loewenson, R., Colvin, C., Rome, N., Nolan, E., Coelho, V., Szabzon, F., Das, S., Aich, U., Tiwari, P., Khanna, R., Gansane, Z., Traoré, Y., Yao, S., Coulibaly, S., Asibu, W., & Chaikosa, S. (2020). *"We Are Subjects, Not Objects in Health": Communities Taking Action on COVID-19*. Training and Research Support Centre in EQUINET and Shapinng Health. https://bit.ly/46kLd9U.

Londsdale, A., Negreiros, P., & Yang, K. (2020). *Urban Climate Finance in the Wake of COVID-19*. Climate Policy Initiative. www.climatepolicyinitiative.org/publication/urban-climate-finance-in-the-wake-of-covid-19/.

Lu, Y., Zhao, J., Wu, X., & Lo, S. M. (2021). Escaping to nature during a pandemic: A natural experiment in Asian cities during the COVID-19 pandemic with big social media data. *Science of the Total Environment*, *777*, 146092. https://doi.org/10.1016/j.scitotenv.2021.146092.

Lührmann, A., Edgell, A. B., & Maerz, S. F. (2020). Pandemic backsliding: Does COVID-19 put democracy at risk? Varieties of Democracy Policy Brief No. 23, Varieties of Democracy (V-Dem) Institute, University of Gothenburg. www.hsdl.org/?abstract&did=842989.

Lwasa, S., & Seto, K. C. (2022). Urban systems and other settlements. In *Climate Change 2022: Mitigation of Climate Change – Contribution of*

Working Group III to the Sixth Assessment Report of the Intergovernmental Panel on Climate Change (pp. 861–952). Cambridge University Press. www.ipcc.ch/report/sixth-assessment-report-working-group-3/.

Mahadevia, D., Adhvaryu, B., Datt, M., & Killiyath, S. (2022). Spatiality of COVID-19 infections in Ahmedabad: An early period analysis. *Urban India*, *42* (1), 1–17.

Mahadevia, D., Pathak, M., Bhatia, N., & Patel, S. (2020). Climate change, heat waves and thermal comfort – reflections on housing policy in India. *Environment and Urbanization ASIA*, *11*(1), 29–50. https://doi.org/10.1177/0975425320906249.

Mahmud, I., & Al-Mohaimeed, A. (2020). COVID-19: Utilizing local experience to suggest optimal global strategies to prevent and control the pandemic. *International Journal of Health Sciences*, *14*(3), 1–3.

Mahtta, R., Mahendra, A., & Seto, K. C. (2019). Building up or spreading out? Typologies of urban growth across 478 cities of 1 million+. *Environmental Research Letters*, *14*(12), 124077. https://doi.org/10.1088/1748-9326/ab59bf.

Marani, M., Katul, G. G., Pan, W. K., & Parolari, A. J. (2021). Intensity and frequency of extreme novel epidemics. *Proceedings of the National Academy of Sciences*, *118*(35), e2105482118. https://doi.org/10.1073/pnas.2105482118.

Marconi, P., Perelman, P., & Salgado, V. (2022). Green in times of COVID-19: Urban green space relevance during the COVID-19 pandemic in Buenos Aires City. *Urban Ecosystems*, *25*, 941–953. https://doi.org/10.1007/s11252-022-01204-z.

Mayor of London – London Assembly. (2024). *Public datasets*. London Datastore – Greater London Authority. https://data.london.gov.uk/dataset/public-transport-journeys-type-transport

McFarlane, C. (2023). Critical commentary: Repopulating density: COVID-19 and the politics of urban value. *Urban Studies*, *60*(9), 1548–1569.

McGowan, V. J., & Bambra, C. (2022). COVID-19 mortality and deprivation: Pandemic, syndemic, and endemic health inequalities. *The Lancet Public Health*, *7*(11), e966–e975. https://doi.org/10.1016/S2468-2667(22)00223-7.

McGuirk, P., Dowling, R., Maalsen, S., & Baker, T. (2021). Urban governance innovation and COVID-19. *Geographical Research*, *59*(2), 188–195. https://doi.org/10.1111/1745-5871.12456.

Melosi, M. V. (2008). *The Sanitary City: Environmental Services in Urban America from Colonial Times to the Present*. University of Pittsburgh Press. https://doi.org/10.2307/j.ctt6wrc97.

Mencos Contreras, E. (2020). Empty Times Square in New York with billboard slogan [Photograph]. Personal Collection.

Mendes, L. (2020). How can we quarantine without a home? Responses of activism and urban social movements in times of COVID-19 pandemic crisis in Lisbon. *Tijdschrift voor Economische en Sociale Geografie* (*Journal of Economic and Human Geography*), *111*(3), 318–332. https://doi.org/10.1111/tesg.12450.

Moctezuma Mendoza, V. (2022). *Pandemia, (pos)neoliberalismo y desamparo del comercio callejero en México. Bitácora Urbano Territorial*, *32*(2), 185–197. https://doi.org/10.15446/bitacora.v32n2.99829.

Mogaji, E. (2020). Impact of COVID-19 on transportation in Lagos, Nigeria. *Transportation Research Interdisciplinary Perspectives*, *6*, 100154. https://doi.org/10.1016/j.trip.2020.100154.

Montag, L., Klünder, T., & Steven, M. (2021). Paving the way for circular supply chains: Conceptualization of a circular supply chain maturity framework. *Frontiers in Sustainability*, *2*. www.frontiersin.org/article/10.3389/frsus.2021.781978.

Moreno, C., Allam, Z., Chabaud, D., Gall, C., & Pratlong, F. (2021). Introducing the "15-minute city": Sustainability, resilience and place identity in future post-pandemic cities. *Smart Cities*, *4*(1), 93–111. https://doi.org/10.3390/smartcities4010006.

Mouratidis, K. (2022). COVID-19 and the compact city: Implications for well-being and sustainable urban planning. *Science of the Total Environment*, *811*, 152332. https://doi.org/10.1016/j.scitotenv.2021.152332.

Murillo, Q. (2022). *Impacto de la COVID 2019 en el comercio informal de los comerciantes regulareizados del mercado Santa Rosa de la ciudad de Riobamba*. Universidad Nacional de Chimbaorazo. http://dspace.unach.edu.ec/handle/51000/9574.

NAA. (2021). Urban migration: Not so fast. National Apartment Association, April 12. www.naahq.org/urban-migration-not-so-fast.

Nastar, M. (2020). Message sent, now what? A critical analysis of the heat action plan in Ahmedabad, India. *Urban Science*, *4*(4), Article 4. https://doi.org/10.3390/urbansci4040053.

NCE. (2016). *The Sustainable Infrastructure Imperative: Financing for Better Growth and Development*. The New Climate Economy. https://bit.ly/3WASWgS.

Nemati, M., & Tran, D. (2022). The Impact of COVID-19 on urban water consumption in the United States. *Water*, *14*(19), 3096. https://doi.org/10.3390/w14193096.

Newell, R., & Dale, A. (2020). COVID-19 and climate change: An integrated perspective. *Cities & Health*, *5*(sup1), S100–S104. https://doi.org/10.1080/23748834.2020.1778844.

Newman, P. (2020). COVID, cities and climate: Historical precedents and potential transitions for the new economy. *Urban Science*, *4*(3), Article 3. https://doi.org/10.3390/urbansci4030032.

Nguyen, S. (2020). Vietnam's unsung recycling heroines have livelihoods ruined by COVID-19. *South China Morning Post*, October 14. https://bit.ly/3VTgRYk.

Nijland, H., & van Meerkerk, J. (2017). Mobility and environmental impacts of car sharing in the Netherlands. *Environmental Innovation and Societal Transitions*, *23*, 84–91. https://doi.org/10.1016/j.eist.2017.02.001.

Niles, M. T., Wirkkala, K. B., Belarmino, E. H., & Bertmann, F. (2021). Home food procurement impacts food security and diet quality during COVID-19. *BMC Public Health*, *21*(1), Article 945. https://doi.org/10.1186/s12889-021-10960-0.

Niu, H., Zhang, C., Hu, W., Hu, T., Wu, C., Hu, S., Silva, L. F. O., Gao, N., Bao, X., & Fan, J. (2022). Air quality changes during the COVID-19 lockdown in an industrial city in North China: Post-pandemic proposals for air quality improvement. *Sustainability*, *14*(18), 11531. https://doi.org/10.3390/su141811531.

Norris, S. A., Frongillo, E. A., Black, M. M., Dong, Y., Fall, C., Lampl, M., Liese, A. D., Naguib, M., Prentice, A., Rochat, T., Stephensen, C. B., Tinago, C. B., Ward, K. A., Wrottesley, S. V., & Patton, G. C. (2022). Nutrition in adolescent growth and development. *The Lancet*, *399*(10320), 172–184. https://doi.org/10.1016/S0140-6736(21)01590-7.

Nunoo, F. (2020). Ghana goment announce free water, 3 month tax holiday give health workers den 50% basic salary give frontline COVID-19 workers. *BBC News Pidgin*, April 6. www.bbc.com/pidgin/tori-52179929.

O'Callaghan, B., & Adam, J.-P. (2021). *Are COVID-19 Fiscal Recovery Measures Bridging or Extending the Emissions Gap?* Emissions Gap Report, UNEP. https://wedocs.unep.org/bitstream/handle/20.500.11822/36996/EGR21_CH5.pdf.

OECD. (2022a). *Global Plastics Outlook: Economic Drivers, Environmental Impacts and Policy Options*. Organisation for Economic Co-operation and Development. www.oecd-ilibrary.org/environment/global-plastics-outlook_de747aef-en.

OECD. (2022b). *Towards a Sustainable Recovery? Carbon Pricing Policy Changes during COVID-19*. Organisation for Economic Co-operation and Development. https://bit.ly/3xhNFkF.

Office for National Statistics. (2020). *Coronavirus (COVID-19) Related Mortality Rates and the Effects of Air Pollution in England*. Office for National Statistics. https://legacy-assets.eenews.net/open_files/assets/2020/08/13/document_gw_07.pdf.

Ogen, Y. (2020). Assessing nitrogen dioxide (NO_2) levels as a contributing factor to coronavirus (COVID-19) fatality. *Science of the Total Environment*, *726*, 138605. https://doi.org/10.1016/j.scitotenv.2020.138605.

Okoh, A. K., Sossou, C., Dangayach, N. S., Meledathu, S., Phillips, O., Raczek, C., Patti, M., Kang, N., Hirji, S. A., Cathcart, C., Engell, C., Cohen, M., Nagarakanti, S., Bishburg, E., & Grewal, H. S. (2020). Coronavirus disease 19 in minority populations of Newark, New Jersey. *International Journal for Equity in Health*, *19*(1), 1–8. https://doi.org/10.1186/s12939-020-01208-1.

Olin, A. (2020). Public transit has lost its momentum during the pandemic. Can it be regained? Kinder Institute for Urban Research, August 5. https://bit.ly/3XzABld.

Onyango, G., & Ondiek, J. O. (2022). Open innovation during the COVID-19 pandemic policy responses in South Africa and Kenya. *Politics & Policy*, *50*(5), 1008–1031. https://doi.org/10.1111/polp.12490.

Paez, A. (2021). Reproducibility of research during COVID-19: Examining the case of population density and the basic reproductive rate from the perspective of spatial analysis. *Geographical Analysis*, *54*(4), 860–880. https://doi.org/10.1111/gean.12307.

Pardo, C. F., Zapata-Bedoya, S., Ramirez-Varela, A., Ramirez-Corrales, D., Espinosa-Oviedo, J.-J., Hidalgo, D., Rojas, N., González-Uribe, C., García, J. D., & Cucunubá, Z. M. (2021). COVID-19 and public transport: An overview and recommendations applicable to Latin America. *Infectio*, *25*(3), 182. https://doi.org/10.22354/in.v25i3.944.

Parreño-Castellano, J., Domínguez-Mujica, J., & Moreno-Medina, C. (2022). Reflections on digital nomadism in Spain during the COVID-19 pandemic – effect of policy and place. *Sustainability*, *14*(23), 16253. https://doi.org/10.3390/su142316253.

Patino, M. (2020). Why Asian countries have succeeded in flattening the curve. *Bloomberg UK*, March 31. www.bloomberg.com/news/articles/2020-03-31/how-to-make-people-stay-home.

Patlins, A. (2021). Adapting the public transport system to the COVID-19 challenge, ensuring its sustainability. *Transportation Research Procedia*, 55. https://trid.trb.org/view/1867947.

Pawar, D. S., Yadav, A. K., Akolekar, N., & Velaga, N. R. (2020). Impact of physical distancing due to novel coronavirus (SARS-CoV-2) on daily travel for work during transition to lockdown. *Transportation Research Interdisciplinary Perspectives*, 7, 100203. https://doi.org/10.1016/j.trip.2020.100203.

Pelling, M., Kerr, R. B., Biesbroek, R., Caretta, M. A., Cissé, G., Costello, M. J., Ebi, K. L., Gunn, E. L., Parmesan, C., Schuster-Wallace, C. J., Tirado, M. C., van Aalst, M., & Woodward, A. (2021). Synergies between COVID-19 and climate change impacts and responses. *Journal of Extreme Events*, 8(3), 2131002. https://doi.org/10.1142/S2345737621310023.

Peng, Y., Wu, P., Schartup, A. T., & Zhang, Y. (2021). Plastic waste release caused by COVID-19 and its fate in the global ocean. *Proceedings of the National Academy of Sciences*, 118(47), e2111530118. https://doi.org/10.1073/pnas.2111530118.

Petrova, M., & Tairov, I. (2022). Solutions to manage smart cities' risks in times of pandemic crisis. *Risks*, 10, 240. https://doi.org/10.3390/risks10120240.

Pisano, C. (2020). Strategies for post-COVID cities: An insight to Paris en Commun and Milano 2020. *Sustainability*, 12(15), 5883. https://doi.org/10.3390/su12155883.

Piscitelli, P., Miani, A., Setti, L., De Gennaro, G., Rodo, X., Artinano, B., Vara, E., Rancan, L., Arias, J., Passarini, F., Barbieri, P., Pallavicini, A., Parente, A., D'Oro, E. C., De Maio, C., Saladino, F., Borelli, M., Colicino, E., Gonçalves, L. M. G., Di Tanna, G., & Domingo, J. L. (2022). The role of outdoor and indoor air quality in the spread of SARS-CoV-2: Overview and recommendations by the research group on COVID-19 and particulate matter (RESCOP commission). *Environmental Research*, 211, 113038. https://doi.org/10.1016/j.envres.2022.113038.

Pitas, N., & Ehmer, C. (2020). Social capital in the response to COVID-19. *American Journal of Health Promotion*, 34(8), 942–944. https://doi.org/10.1177/0890117120924531.

PNUMA. (2021). *El Peso de las Ciudades en América Latina y el Caribe: Requerimientos Futuros de Recursos y Potenciales Rutas de Actuación*. Programa de las Naciones Unidas para el Medio Ambiente, Oficina para América Latina y el Caribe. https://bit.ly/3LEHnPh.

Prater, E., & Lichtenberg, N. (2022). The pandemic migration's full impact is becoming clear – and it's a "big deal" for the future of cities and white-collar work. *Fortune*, April 3. https://bit.ly/45Rftct.

Prieur-Richard, A.-H., Walsh, B., Craig, M., Melamed, M. L., Pathak, M., Connors, S., Bai, X., Barau, A., Bulkeley, H., Cleugh, H., Cohen, M., Colenbrander, S., Dodman, D., Dhakal, S., Dawson, R., Greenwalt, J., Kurian, P., Lee, B., Leonardsen, L., Masson-Delmotte, V., Munshi, D., Okem, A., Delgado Ramos, G. C., Sanchez Rodriguez, R., Roberts, D., Rosenzweig, C., Schultz, S., Seto, K., Solecki, W., van Staden, M., & Ürge-Vorsatz, D. (2018). *Extended version: Global Research and Action Agenda on Cities and Climate Change Science*. WCRP Publication.

Pulighe, G., & Lupia, F. (2020). Food first: COVID-19 outbreak and cities lockdown a booster for a wider vision on urban agriculture. *Sustainability*, *12*(12), Article 12. https://doi.org/10.3390/su12125012.

PWC & Urban Land Institute. (2022). *Emerging Trends in Real Estate 2023*. https://bit.ly/3KVsPKW.

Qiu, R. T. R., Park, J., Li, S., & Song, H. (2020). Social costs of tourism during the COVID-19 pandemic. *Annals of Tourism Research*, *84*, 102994. https://doi.org/10.1016/j.annals.2020.102994.

Rahman, S. M. M., Kim, J., & Laratte, B. (2021). Disruption in circularity? Impact analysis of COVID-19 on ship recycling using Weibull tonnage estimation and scenario analysis method. *Resources, Conservation and Recycling*, *164*, 105139. https://doi.org/10.1016/j.resconrec.2020.105139.

Rajan, S. I., & Cherian, A. P. (2021). COVID-19: Urban vulnerability and the need for transformations. *Environment and Urbanization ASIA*, *12*(2), 310–322. https://doi.org/10.1177/09754253211040195.

Revi, A., Satterthwaite, D. E., Aragón-Durand, F., Corfee-Morlot, J., Kiunsi, R. B. R., Pelling, M., Roberts, D. C., & Solecki, W. (2014). Urban areas. In *Climate change 2014: Impacts, Adaptation, and Vulnerability – Part A: Global and Sectoral Aspects (Contribution of Working Group II to the Fifth Assessment Report of the Intergovernmental Panel on Climate Change)*. Cambridge University Press.

Ribeiro, H. V., Sunahara, A. S., Sutton, J., Perc, M., & Hanley, Q. S. (2020). City size and the spreading of COVID-19 in Brazil. *PLOS ONE*, *15*(9), e0239699. https://doi.org/10.1371/journal.pone.0239699.

Roberts, B. H., & Drake, J. (Eds.). (2021). *Secondary Cities Post-COVID 19: Achieving Urban Sustainable and Regenerative Development in Emerging Economies*. Cities Alliance.

Rogerson, C. M., & Rogerson, J. M. (2020). COVID-19 and tourism spaces of vulnerability in South Africa. *African Journal of Hospitality, Tourism and Leisure*, *9*(4), 382–401. www.ajhtl.com/uploads/7/1/6/3/7163688/article_1_9_4___382-401.pdf.

Romanello, M., McGushin, A., Napoli, C. D., Drummond, P., Hughes, N., Jamart, L., Kennard, H., Lampard, P., Rodriguez, B. S., Arnell, N., Ayeb-Karlsson, S., Belesova, K., Cai, W., Campbell-Lendrum, D., Capstick, S., Chambers, J., Chu, L., Ciampi, L., Dalin, C., . . . Hamilton, I. (2021). The 2021 report of the *Lancet* Countdown on health and climate change: Code red for a healthy future. *The Lancet*, *398*(10311), 1619–1662. https://doi.org/10.1016/S0140-6736(21)01787-6.

Rosenzweig, C., Solecki, W., Hammer, S.A., & Mehrotra, S. (Eds.). (2011). *Climate Change and Cities: First Assessment Report of the Urban Climate Change Research Network*. Cambridge University Press.

Rosenzweig, C., Solecki, W., Romero-Lankao, P., Mehrotra, S., Dhakal, S., & Ali Ibrahim, S. (Eds.). (2018). *Climate Change and Cities: Second Assessment Report of the Urban Climate Change Research Network*. Cambridge University Press.

Rouleau, J., & Gosselin, L. (2021). Impacts of the COVID-19 lockdown on energy consumption in a Canadian social housing building. *Applied Energy*, *287*, 116565. https://doi.org/10.1016/j.apenergy.2021.116565.

Rowe, B. R., Canosa, A., Drouffe, J. M., & Mitchell, J. B. A. (2021). Simple quantitative assessment of the outdoor versus indoor airborne transmission of viruses and COVID-19. *Environmental Research*, *198*, 111189. https://doi.org/10.1016/j.envres.2021.111189.

Roy, P., Mohanty, A. K., Wagner, A., Sharif, S., Khalil, H., & Misra, M. (2021). Impacts of COVID-19 outbreak on the municipal solid waste management: Now and beyond the pandemic. *ACS Environmental Au*, *1*(1), 32–45. https://doi.org/10.1021/acsenvironau.1c00005.

Russette, H., Graham, J., Holden, Z., Semmens, E. O., Williams, E., & Landguth, E. L. (2021). Greenspace exposure and COVID-19 mortality in the United States: January–July 2020. *Environmental Research*, *198*, 111195. https://doi.org/10.1016/j.envres.2021.111195.

Ruszczyk, H. A., Rahman, M. F., Bracken, L. J., & Sudha, S. (2021). Contextualizing the COVID-19 pandemic's impact on food security in two small cities in Bangladesh. *Environment and Urbanization*, *33*(1), 239–254. https://doi.org/10.1177/0956247820965156.

Sadanandan, R. (2020). Kerala's response to COVID-19. *Indian Journal of Public Health*, *64*(6), 99–101. https://doi.org/10.4103/ijph.IJPH_459_20.

Sah, R., Sigdel, S., Ozaki, A., Kotera, Y., Bhandari, D., Regmi, P., Rabaan, A. A., Mehta, R., Adhikari, M., Roy, N., Dhama, K., Tanimoto, T., Rodríguez-Morales, A. J., & Dhakal, R. (2020). Impact of COVID-19 on tourism in Nepal. *Journal of Travel Medicine*, *27*(6), taaa105. https://doi.org/10.1093/jtm/taaa105.

Sasikumar, K., Nath, D., Nath, R., & Chen, W. (2020). Impact of extreme hot climate on COVID-19 outbreak in India. *GeoHealth*, *4*(12), e2020GH000305. https://doi.org/10.1029/2020GH000305.

Satterthwaite, D., Archer, D., Colenbrander, S., Dodman, D., Hardoy, J., & Patel, S. (2018). Responding to climate change in cities and in their informal settlements and economies. Paper prepared for the IPCC for the International Scientific Conference on Cities and Climate Change in Edmonton. International Institute for Enviroment and Development. www.iied.org/sites/default/files/pdfs/migrate/G04328.pdf.

SEDEMA. (2016). *Suelo de Conservación*. www.sedema.cdmx.gob.mx/storage/app/media/Libro_Suelo_de_Conservacion.pdf.

SEDEMA. (2022). *Cuarto Informe de Gobierno de la Ciudad de México, 2019–2022*. https://bit.ly/3yefD0W.

Sempewo, J. I., Kisaakye, P., Mushomi, J., Tumutungire, M. D., & Ekyalimpa, R. (2021). Assessing willingness to pay for water during the COVID-19 crisis in Ugandan households. *Social Sciences & Humanities Open*, *4*(1), 100230. https://doi.org/10.1016/j.ssaho.2021.100230.

Serrano, A., & Gutierrez Torres, D. (2020). Latin America moving fast to ensure water services during COVID-19. World Bank Blogs, April 8. https://bit.ly/4cuBBeD.

Seto, K. C., Dhakal, S., Bigio, A., Blanco, H., Delgado Ramos, G. C., Dewar, D., Huang, L., Inaba, A., Kansal, A., Lwasa, S., McMahon, J., Müller, D. B., Murakami, J., Nagendra, H., & Ramaswami, A. (2014). Human settlements, infrastructure, and spatial planning. In *Climate Change 2014*: *Mitigation of Climate Change: Contribution of Working Group III to the Fifth Assessment Report of the Intergovernmental Panel on Climate Change* (pp. 923–1000). Cambridge University Press. https://doi.org/10.1017/CBO9781107415416.

Sharifi, A., & Khavarian-Garmsir, A. R. (2020). The COVID-19 pandemic: Impacts on cities and major lessons for urban planning, design, and management. *Science of the Total Environment*, *749*, 142391. https://doi.org/10.1016/j.scitotenv.2020.142391.

Sharma, H. B., Vanapalli, K. R., Samal, B., Cheela, V. R. S., Dubey, B. K., & Bhattacharya, J. (2021). Circular economy approach in solid waste management system to achieve UN-SDGs: Solutions for post-COVID recovery. *Science of the Total Environment*, *800*, 149605. https://doi.org/10.1016/j.scitotenv.2021.149605.

References 97

Sharma, K.V. [@ikaransharma27]. (2020). *COVID-19 testing in Dharavi*. [Photograph]. Twitter. Retrieved August 14, 2024.

Shultz, J. M., Fugate, C., & Galea, S. (2020). Cascading risks of COVID-19 resurgence during an active 2020 Atlantic hurricane season. *JAMA*, *324*(10), 935–936. https://doi.org/10.1001/jama.2020.15398.

Simon, C. (2022). Portland's 20-minute neighborhoods after ten years: How a planning initiative impacted accessibility. Unpublished master's thesis, University of Washington. https://bit.ly/3VTapR8.

Simonovic, S. P., Kundzewicz, Z. W., & Wright, N. (2021). Floods and the COVID-19 pandemic: A new double hazard problem. *WIREs Water*, *8*(2), e1509. https://doi.org/10.1002/wat2.1509.

Šinko, S., Prah, K., & Kramberger, T. (2021). Spatial modelling of modal shift due to COVID-19. *Sustainability*, *13*(13), 7116. https://doi.org/10.3390/su13137116.

Sneader, K., & Lund, S. (2020). COVID-19 and climate change expose dangers of unstable supply chains. McKinsey & Company, August 18. https://bit.ly/3zbs2mt.

Soga, M., Evans, M. J., Cox, D. T. C., & Gaston, K. J. (2021). Impacts of the COVID-19 pandemic on human–nature interactions: Pathways, evidence and implications. *People and Nature*, *3*(3), 518–527. https://doi.org/10.1002/pan3.10201.

Solecki, W., Delgado Ramos, G. C., Roberts, D., Rosenzweig, C., & Walsh, B. (2021). Accelerating climate research and action in cities through advanced science–policy–practice partnerships. *npj Urban Sustainability*, *1*(1), Article 1. https://doi.org/10.1038/s42949-021-00015-z.

Solecki, W., Pathak, M., Barata, M., Salisu Barau, A., Dombrov, M., & Rosenzweig, C. (Eds.). (in press). *Climate Change and Cities: Third Assessment Report of the Urban Climate Change Research Network*. Cambridge University Press.

Spear, R., Erdi, G., Parker, M. A., & Anastasiadis, M. (2020). Innovations in citizen response to crises: Volunteerism and social mobilization during COVID-19. *Interface: A Journal for and about Social Movements*, *12*(1), 383–391. https://bit.ly/4bftWQg.

Spooner, D., & Whelligan, J. (2020). *Informal Passenger Transport beyond COVID-19: A Trade Union Guide to Worker-Led Formalisation*. Our Public Transport. https://bit.ly/45Xbrzb.

Suárez Lastra, M., Valdés González, C., Galindo Pérez, C., Salvador Guzmán, E., Ruiz Rivera, N., Alcántara-Ayala, I., Lopez-Cervantes, M., Rosales Tapia, A., Lee, W., Benítez-Pérez, H., Juárez Gutiérrez, M. del C., Bringas López, A., Oropeza Orozco, O., Peralta Higuera, A., & Garnica

Peña, R. (2021). Índice de vulnerabilidad ante el COVID-19 en México. *Investigaciones Geográficas*, *104*, e60140. https://dx.doi.org/10.14350/rig.60140.

Surahman, U., Hartono, D., Setyowati, E., & Jurizat, A. (2022). Investigation on household energy consumption of urban residential buildings in major cities of Indonesia during COVID-19 pandemic. *Energy and Buildings*, *261*, 111956. https://doi.org/10.1016/j.enbuild.2022.111956.

Sverdlik, A., & Walnycki, A. (2021). *Better Cities after COVID-19: Transformative Urban Recovery in the Global South*. International Institute for Environment and Development. https://pubs.iied.org/sites/default/files/pdfs/2021-06/20241iied.pdf.

Swarna, N. R., Anjum, I., Hamid, N. N., Rabbi, G. A., Islam, T., Evana, E. T., Islam, N., Rayhan, Md. I., Morshed, K., & Miah, A. S. Md. J. (2022). Understanding the impact of COVID-19 on the informal sector workers in Bangladesh. *PLOS ONE*, *17*(3), e0266014. https://doi.org/10.1371/journal.pone.0266014.

Tabibzadeh, A., Esghaei, M., Soltani, S., Yousefi, P., Taherizadeh, M., Tameshkel, F. S., Golahdooz, M., Panahi, M., Ajdarkosh, H., Zamani, F., & Niya, M. H. K. (2021). Evolutionary study of COVID-19, severe acute respiratory syndrome coronavirus 2 (SARS-CoV-2) as an emerging coronavirus: Phylogenetic analysis and literature review. *Veterinary Medicine and Science*, *7*(2), 559–571. https://doi.org/10.1002/vms3.394.

Tageo, V., Dantas, C., Corsello, A., & Dias, L. (2021). *The Response of Civil Society Organisations to Face the COVID-19 Pandemic and the Consequent Restrictive Measures Adopted in Europe*. European Economic and Social Committee. www.eesc.europa.eu/sites/default/files/files/qe-02-21-011-en-n.pdf.

Tayal, S., & Singh, S. (2021). Covid-19 and opportunity for integrated management of water–energy–food resources for urban consumption. In A. L. Ramanathan, C. Sabarathinam, F. Arriola, M. V. Prasanna, P. Kumar, & M. P. Jonathan (Eds.), *Environmental Resilience and Transformation in Times of COVID-19* (pp. 135–142). Elsevier. https://bit.ly/3zh7aKI.

Taylor, J. (2020). How Dhaka's urban poor are dealing with COVID-19. International Institute for Environment and Development, July 1. www.iied.org/how-dhakas-urban-poor-are-dealing-covid-19.

Teller, J. (2021). Urban density and Covid-19: Towards an adaptive approach. *Buildings and Cities*, *2*(1), 150–165.

Termeer, C. J. A. M., Dewulf, A., & Biesbroek, R. (2019). A critical assessment of the wicked problem concept: Relevance and usefulness for policy science

and practice. *Policy and Society*, *38*(2), 167–179. https://doi.org/10.1080/14494035.2019.1617971.
TfL. (2020). *Transport for London: Green Bond Framework 2020*. Transport For London. https://content.tfl.gov.uk/tfl-green-bond-framework-2020.pdf.
Thombre, A., & Agarwal, A. (2021). A paradigm shift in urban mobility: Policy insights from travel before and after COVID-19 to seize the opportunity. *Transport Policy*, *110*, 335–353. https://doi.org/10.1016/j.tranpol.2021.06.010.
UCLG, Metropolis, & LSE. (2021). Multivel governance and COVID-19 emergency coordination: Emergency governance for cities annd regions. Analytics Note #04, Emergency Governance Initiative. https://bit.ly/3zngT1S.
Ugolini, F., Massetti, L., Calaza-Martínez, P., Cariñanos, P., Dobbs, C., Ostoić, S. K., Marin, A. M., Pearlmutter, D., Saaroni, H., Šaulienė, I., Simoneti, M., Verlič, A., Vuletić, D., & Sanesi, G. (2020). Effects of the COVID-19 pandemic on the use and perceptions of urban green space: An international exploratory study. *Urban Forestry & Urban Greening*, *56*, 126888. https://doi.org/10.1016/j.ufug.2020.126888.
UN Women. (2021). How six grass-roots women's organizations are making sure that no one is left behind in COVID-19 response. UN Women, March 10. https://bit.ly/3RGvsUz.
UNDESA. (2019). *World Urbanization Prospects: The 2018 Revision*. United Nations Department of Economic and Social Affairs. https://population.un.org/wup/Publications/Files/WUP2018-Report.pdf.
UNDESA. (2022). *E-Government Survey 2022: The Future of Digital Government*. United Nations Deparment of Economic and Social Affairs. https://bit.ly/3VEaOWB.
UNDP. (2022). *New Threats to Human Security in the Anthropocene: Demanding Greater Solidarity*. United Nations Development Programme. https://bit.ly/45GEzdJ.
UNEP. (2021). *The Heat Is On: A World of Climate Promises Not Yet Delivered*. Emissions Gap Report 2021, United Nations Environment Programme. www.unep.org/es/resources/emissions-gap-report-2021.
UNEP & UN-Habitat. (2021). *Global Environment Outlook for Cities: Towards Green and Just Cities*. United Nations Environment Programme and United Nations Human Settlements Programme. https://wedocs.unep.org/bitstream/handle/20.500.11822/37413/GEOcities.pdf.
UN-Habitat. (2020a). *UN-Habitat COVID-19 Response Plan*. https://bit.ly/4fiN2bA.

UN-Habitat. (2020b). Urban transport and COVID-19: Key messages. https://bit.ly/4bg8NW8.

UN-Habitat. (2021). *Life amidst a Pandemic: Urban Livelihoods, Food Security And Nutrition in Sub-Saharan Africa.* https://bit.ly/4cQSuAc.

UN-Habitat. (2022). Food insecurity a real concern among the urban poor in sub-saharan Africa following pandemic – new report shows. News Release, March 11. https://bit.ly/4cAM9ZU.

United Nations Human Settlements Programme. (2022). *UN-Habitat.* https://unhabitat.citiiq.com

UNU-WIDER & WIEGO. (2022). COVID-19 and informal work: Degrees and pathways of impact in 11 cities around the world. WIDER Working Paper 2022/45, United Nations University World Institute for Development Economics Research. https://bit.ly/3VARrxw.

Unwin, H. J. T., Hillis, S., Cluver, L., Flaxman, S., Goldman, P. S., Butchart, A., Bachman, G., Rawlings, L., Donnelly, C. A., Ratmann, O., Green, P., Nelson, C. A., Blenkinsop, A., Bhatt, S., Desmond, C., Villaveces, A., & Sherr, L. (2022). Global, regional, and national minimum estimates of children affected by COVID-19-associated orphanhood and caregiver death, by age and family circumstance up to Oct 31, 2021: An updated modelling study. *The Lancet Child & Adolescent Health*, 6(4), 249–259. https://doi.org/10.1016/S2352-4642(22)00005-0.

US Census Bureau. (2022). Over two-thirds of the nation's counties had natural decrease in 2021. Press Release, March 24. https://bit.ly/3RGAZL7.

Valdés, J. E. V., & Caballero, C. V. R. (2020). Air pollution and mobility in the Mexico City Metropolitan Area in times of COVID-19. *Atmósfera*, 36(2), 343–354. https://doi.org/10.20937/ATM.53052.

Valenzuela-Levi, N., Echiburu, T., Correa, J., Hurtubia, R., & Muñoz, J. C. (2021). Housing and accessibility after the COVID-19 pandemic: Rebuilding for resilience, equity and sustainable mobility. *Transport Policy*, 109, 48–60. https://doi.org/10.1016/j.tranpol.2021.05.006.

Vasquez-Apestegui, B. V., Parras-Garrido, E., Tapia, V., Paz-Aparicio, V. M., Rojas, J. P., Sánchez-Ccoyllo, O. R., & Gonzales, G. F. (2021). Association between air pollution in Lima and the high incidence of COVID-19: Findings from a post hoc analysis. *BMC Public Health.* https://doi.org/10.21203/rs.3.rs-39404/v2.

Vega, E., Namdeo, A., Bramwell, L., Miquelajauregui, Y., Resendiz-Martinez, C. G., Jaimes-Palomera, M., Luna-Falfan, F., Terrazas-Ahumada, A., Maji, K. J., Entwistle, J., Enríquez, J. C. N., Mejia, J. M., Portas, A., Hayes, L., & McNally, R. (2021). Changes in air quality in Mexico City, London and Delhi in response to various stages and levels of lockdowns and easing of restrictions during COVID-19 pandemic. *Environmental Pollution*, *285*, 117664. https://doi.org/10.1016/j.envpol.2021.117664.

Venter, Z. S., Barton, D. N., Gundersen, V., Figari, H., & Nowell, M. (2020). Urban nature in a time of crisis: Recreational use of green space increases during the COVID-19 outbreak in Oslo, Norway. *Environmental Research Letters*, *15*(10), 104075. https://doi.org/10.1088/1748-9326/abb396.

Victoria State Government. (2019). *Sunshine West: Our 20-Minute Neighbourhood*. https://bit.ly/3SoDW3e.

Villeneuve, H., Abdeen, A., Papineau, M., Simon, S., Cruickshank, C., & O'Brien, W. (2021). New Insights on the energy impacts of telework in Canada. *Canadian Public Policy*, *47*(3), 460–477. https://doi.org/10.3138/cpp.2020-157.

Vitrano, C. (2021). *COVID-19 and Public Transport. A Review of the International Academic Literature*. https://bit.ly/3WhVXRX.

Wahid, W. W. C., & Setyono, J. S. (2022). The urban environment and public health: Associations between COVID-19 cases and urban factors in Semarang City, Central Java, Indonesia. *IOP Conference Series: Earth and Environmental Science*, *1111*(1), 012067. https://doi.org/10.1088/1755-1315/1111/1/012067.

Walton, D., Arrighi, J., van Aalst, M., & Claudet, M. (2021). *The Compound Impact of Extreme Weather Events and COVID-19: An Update of the Number of People Affected and a Look at the Humanitarian Implications in Selected Contexts*. International Federation of Red Cross and Red Crescent Societies. https://bit.ly/3SpM0AH.

Wang, B., Liu, J., Fu, S., Xu, X., Li, L., Ma, Y., Zhou, J., Yaoc, J., Liu, X., Zhang, X., He, X., Yan, J., Shi, Y., Ren, X., Niu, J., Luo, B., & Zhang, K. (2020). Airborne particulate matter, population mobility and COVID-19: A multi-city study in China. *MC Public Health*, *20*, 1585. https://doi.org/10.1186/s12889-020-09669-3.

Ware, G., & Mariwany, M. (2022). When digital nomads come to town: Governments want their cash but locals are being left behind – podcast. *The Conversation*, Ocotober 20. https://bit.ly/4et2Xn8.

Watkins, G., Breton, H., & Edwards, G. (2021). *Achieving Sustainable Recovery: Criteria for Evaluating the Sustainability and Effectiveness of Covid-19 Recovery Investments in Latin America and the Caribbean*. Inter-American Development Bank. http://dx.doi.org/10.18235/0003413.

WCRP. (2019). *Global Research and Action Agenda on Cities and Climate Change Science*. World Climate Research Programme. www.wcrp-climate.org/news/wcrp-news/1517-graa-published.

WEF. (2022). *Using Digital Technology for a Green and Just Recovery in Cities*. World Economic Forum. www3.weforum.org/docs/WEF_C4IR_GFC_on_Cities_Digital_Technology_2022.pdf.

Whitaker, S. D. (2021). *Did the COVID-19 Pandemic Cause an Urban Exodus?* Cleveland Fed District Data Brief. https://bit.ly/3zccqyZ.

WHO. (2016). *Ambient Air Pollution: A Global Assessment of Exposure and Burden of Disease*. World Health Organization. https://apps.who.int/iris/handle/10665/250141.

WHO. (2020a). Recommendations to member states to improve hand hygiene practices to help prevent the transmission of the COVID-19 virus. World Health Organization. https://bit.ly/45BiuNO.

WHO. (2020b). *Report of the WHO–China Joint Mission on Coronavirus Disease 2019 (COVID-19)*. World Health Organization. https://bit.ly/3RF8ZHJ.

WHO. (2020c). Listings of WHO's response to COVID-19. Website. www.who.int/news/item/29-06-2020-covidtimeline.

WHO. (2021a). *COP26 Special Report on Climate Change and Health: The Health Argument for Climate Action*. World Health Organization. https://apps.who.int/iris/handle/10665/346168.

WHO. (2021b). *Roadmap to Improve and Ensure Good Indoor Ventilation in the Context of COVID-19*. World Health Organization. https://apps.who.int/iris/handle/10665/339857.

WHO. (2023). Coronavirus (COVID-19) dashboard. Website. https://covid19.who.int.

WIEGO. (2020). *Informal Workers in the COVID-19 Crisis: A Global Picture of Sudden Impact and Long-Term Risk*. Women in Informal Employment: Globalizing and Organizing. https://bit.ly/3xCJekg.

WIEGO. (2022). *COVID-19 and Informal Work in 11 Cities: Recovery Pathways amidst Continued Crisis*. Women in Informal Employment: Globalizing and Organizing. https://bit.ly/4byFM8o.

Woodward, A., Smith, K. R., Campbell-Lendrum, D., Chadee. D. D., Honda, Y., Liu, Q., Olwoch, J., Revich, B., Sauerborn, R., Chafe, Z., Confalonieri, U., & Haines, A. (2014). Climate change and health: On The latest IPCC report. *The Lancet*, *383*(9924), 1185–1189. https://doi.org/10.1016/S0140-6736(14)60576-6.

Wu, X., Nethery, R. C., Sabath, M. B., Braun, D., & Dominici, F. (2020). Air pollution and COVID-19 mortality in the United States: Strengths and limitations of an ecological regression analysis. *Science Advances*, *6*(45), eabd4049. https://doi.org/10.1126/sciadv.abd4049.

Xinhua. (2020). Thai gov't grants free tap water for households during stay-home period to counter COVID-19 pandemic. *Xinhua*, May 5. www.xinhuanet.com/english/asiapacific/2020-05/05/c_139032933.htm.

Xu, Y., & Juan, Y.-K. (2021). Design strategies for multi-unit residential buildings during the post-pandemic era in China. *Frontiers in Public Health*, *9*. www.frontiersin.org/articles/10.3389/fpubh.2021.761614.

Yang, Y., Cao, M., Cheng, L., Zhai, K., Zhao, X., & De Vos, J. (2021). Exploring the relationship between the COVID-19 pandemic and changes in travel behaviour: A qualitative study. *Transportation Research Interdisciplinary Perspectives*, *11*, 100450. https://doi.org/10.1016/j.trip.2021.100450.

Yao, Y., Pan, J., Liu, Z., Meng, X., Wang, W., Kan, H., & Wang, W. (2021). Ambient nitrogen dioxide pollution and spreadability of COVID-19 in Chinese cities. *Ecotoxicology and Environmental Safety*, *208*, 111421. https://doi.org/10.1016/j.ecoenv.2020.111421.

Yassin, H. H. (2019). Livable city: An approach to pedestrianization through tactical urbanism. *Alexandria Engineering Journal*, *58*(1), 251–259. https://doi.org/10.1016/j.aej.2019.02.005.

Yokelson, R. J., Urbanski, S. P., Atlas, E. L., Toohey, D. W., Alvarado, E. C., Crounse, J. D., Wennberg, P. O., Fisher, M. E., Wold, C. E., Campos, T. L., Adachi, K., Buseck, P. R., & Hao, W. M. (2007). Emissions from forest fires near Mexico City. *Atmospheric Chemistry and Physics*, *7*, 5569–5584. https://doi.org/10.5194/acp-7-5569-2007.

Zhang, J., Hayashi, Y., & Frank, L. D. (2021). COVID-19 and transport: Findings from a world-wide expert survey. *Transport Policy*, *103*, 68–85. https://doi.org/10.1016/j.tranpol.2021.01.011.

Acknowledgments

We thank UCCRN Hub Directors – Minal Pathak (South Asia Hub), Ruishan Chen (East Asia Hub), and Chantal Pacteau (Europe Hub) – for their excellent shepherding of this Element. Jaad Benhallam and Natalie Kozlowski of the UCCRN Secretariat provided outstanding editorial and graphic support. We greatly appreciate the constructive reviews of the Element by Hilda Blanco, Renee van Diemen, Kris Ebi, Deljana Iossifova, Anjali Mahendra, and Dustin Robertson. UCCRN acknowledges support from NASA (WBS 509496.02.80.01.16) for this publication.

Author List

Coordinating Lead Authors

Darshini Mahadevia, *Ahmedabad University, Ahmedabad*
Gian C. Delgado Ramos, *National Autonomous University of Mexico, Mexico City*

Lead Authors

Janice Barnes, *Climate Adaptation Partners and the University of Pennsylvania, New York/Philadelphia*
Joan Fitzgerald, *Northeastern University, Boston*
Miho Kamei, *Institute for Global Environmental Strategies, Hayama*
Kevin Lanza, *University of Texas Health Science Center at Houston (UTHealth), Austin*

Contributing/Case Study Authors

Zaheer Allam, *University of Paris, Paris*
Amita Bhide, *Tata Institute of Social Sciences, Mumbai*
Yakubu Bununu, *Ahmadu Bello University, Zaria*
Didier Chabaud, *University of Paris, Paris*
Amitkumar Dubey, *Ahmedabad University, Ahmedabad*
Yann Francoise, *Climate and Ecological Transition Directorate at Paris City, Paris*
Saumya Lathia, *Ahmedabad University, Ahmedebad*
María Fernanda Mac Gregor-Gaona, *National Autonomous University of Mexico, Mexico City*
Carlos Moreno, *Paris 1 Sorbonne University, Tunja/Paris*
Marie-Christine Therrien, *Ecole Nationale d'Administration Publique – Cité-ID LivingLab Urban Resilience Governance, Montreal*
Nada Toueir, *Lincoln University, Montreal*
Nelzair Vianna, *Oswaldo Cruz Foundation, Salvador*

Element Scientist

Melissa López Portillo-Purata, *National Autonomous University of Mexico, Mexico City*

Cambridge Elements

Climate Change and Cities: Third Assessment Report of the Urban Climate Change Research Network

Series Editors

William Solecki
New York

William Solecki is a Professor in the Department of Geography at Hunter College, City University of New York (CUNY). From 2006–2014 he served as the Director of the CUNY Institute for Sustainable Cities at Hunter College. He also served as interim Director of the Science and Resilience Institute at Jamaica Bay. He has co-led several climate impacts studies in the greater New York and New Jersey region, including the New York City Panel on Climate Change (NPCC). He was a lead author of the Urban Areas chapter in the *Fifth Assessment Report of the Intergovernmental Panel on Climate Change* (IPCC), and a coordinating lead author of the Urbanization, Infrastructure, and Vulnerability chapter in the *Third National Climate Assessment (US)*. He is a co-founder of the Urban Climate Change Research Network (UCCRN), co-editor of *Current Opinion in Environmental Sustainability*, and founding editor of the *Journal of Extreme Events*. His research focuses on urban environmental change, resilience, and adaptation transitions.

Minal Pathak
Ahmedabad

Minal Pathak is an Associate Professor at the Global Centre for Environment and Energy at Ahmedabad University, India. She is a Senior Scientist with the Technical Support Unit of Working Group III of the IPCC for its Sixth Assessment cycle. She has contributed to two IPCC special reports, co-edited the IPCC *Sixth Assessment Report*, and contributed to the recently published IPCC *Sixth Assessment Synthesis Report*. She heads the South Asia Hub of the UCCRN, headquartered at the Columbia Climate School. She was a visiting researcher at Imperial College London (2017–2023) and a visiting scholar at MIT (2016–2017). Her research focuses on climate change mitigation strategies for urban settlements, transport, and buildings and their co-benefits/interlinkages with development.

Martha Barata
Rio de Janeiro

Martha Barata is Coordinator of the Latin America Hub of the UCCRN, headquartered at Columbia Climate School. Barata is a collaborating researcher at the Oswaldo Cruz Institute (Fiocruz) and CentroClima (COPPE/UFRJ), following retirement from the Oswaldo Cruz Foundation in 2017. She was a visiting scholar in the Center for Climate Systems Research at Columbia University in 2014, conducting research on climate risk management in cities.

Aliyu Salisu Barau
Kano

Aliyu Salisu Barau is a Professor in Urban Development and Management at the Department of Urban and Regional Planning and Fifth Dean of the Faculty of Earth and

Environmental Sciences at Bayero University in Kano, Nigeria. He is a transdisciplinary researcher with interests in climate change, landscape ecology, clean energy, socio-ecological systems, sustainability agenda setting, informally and formally protected eco-systems, special economic zones, and inclusive and innovative planning. He contributes to the research, policy, and action agenda in Nigeria and globally through engagements with UN Environment, IPCC, Future Earth, IUCN, IPBES, IIED, UNICEF, and UN-Habitat. He is also the director of the West Africa Center for the UCCRN at Columbia University in New York.

Maria Dombrov
New York

Maria Dombrov is a Senior Research Associate I at the Climate Impacts Group, co-located at NASA Goddard Institute for Space Studies and Columbia University's Center for Climate Systems Research in New York City. Ms. Dombrov is UCCRN's Global Coordinator and the Project Manager of UCCRN's *Third Assessment Report on Climate Change and Cities* (ARC3.3). Her work focuses on understanding the risks and vulnerabilities that climate change and extreme events present to cities and their metropolitan regions around the world.

Cynthia Rosenzweig
New York

Cynthia Rosenzweig is a Senior Research Scientist at the NASA Goddard Institute for Space Studies (GISS), Adjunct Senior Research Scientist at the Columbia University Climate School, and Adjunct Professor in the Department of Environmental Science at Barnard College. At NASA GISS, she heads the Climate Impacts Group, whose mission is to investigate the interactions of climate on systems and sectors important to human well-being.
Dr. Rosenzweig is Co-Founder and Co-Director of the Urban Climate Change Research Network (UCCRN). She was Co-Chair of the New York City Panel on Climate Change (NPCC). Dr. Rosenzweig has served as Coordinating Lead Author and Lead Author on several IPCC Assessment Reports.

About the Series

This Elements series, published in collaboration with the Urban Climate Change Research Network (UCCRN), provides essential knowledge on climate change and cities for researchers, practitioners, policymakers, and students. Bridging the gap between theory and practice, the series invites readers to engage with the latest advances in the field. Focusing on urban transformation, cross-cutting themes, and urgent research areas, it empowers stakeholders to drive impactful climate action in rapidly-evolving urban contexts.

Cambridge Elements

Climate Change and Cities: Third Assessment Report of the Urban Climate Change Research Network

Elements in the Series

Learning from COVID-19 for Climate-Ready Urban Transformation
Darshini Mahadevia, Gian C. Delgado Ramos, Janice Barnes, Joan Fitzgerald, Miho Kamei, and Kevin Lanza

A full series listing is available at: www.cambridge.org/ECCC